Smart Agents for the Industry 4.0

Max Hoffmann

Smart Agents for the Industry 4.0

Enabling Machine Learning in Industrial Production

Max Hoffmann
Institute of Information Management in
Mechanical Engineering
RWTH Aachen University
Aachen, Germany

Dissertation, RWTH Aachen University, 2017, under the title "Adaptive and Scalable Information Modeling to Enable Autonomous Decision Making for Real-Time Interoperable Factories"

ISBN 978-3-658-27744-4   ISBN 978-3-658-27742-0   (eBook)
https://doi.org/10.1007/978-3-658-27742-0

Springer Vieweg
© Springer Fachmedien Wiesbaden GmbH, part of Springer Nature 2019

This Springer Vieweg imprint is published by the registered company Springer Fachmedien Wiesbaden GmbH part of Springer Nature.
The registered company address is: Abraham-Lincoln-Str. 46, 65189 Wiesbaden, Germany

Foreword from Academia

The subject of Max Hoffmann's dissertation at the Institute for Information Management in Mechanical Engineering (IMA) at RWTH Aachen University is to investigate, deepen and offer solutions for a still existing central challenge on the way to the Smart Factory. The thesis deals with the interoperability between machines and robots, especially the consistent representation of manufacturing resources by a formal, common information model.

Modern industrial production plants and their peripheral processes are characterized by changing framework conditions. These may be necessary modifications due to changes in the product portfolio, malfunctions or supply bottlenecks, as well as the introduction and expansion of digital interfaces and services. Modern production is becoming ever more digitalized and data-intensive. Nevertheless, classic approaches are not suited to manage these changes. One chance to meet these challenges is to decentralize the control mechanisms. Multi-agent systems and distributed artificial intelligence provide an excellent starting point for this.

Mr. Max Hoffmann has described and developed a reference architecture in which autonomous units are able to plan and track independent tasks in a production process considering their "personal" environmental conditions. The reference architecture described by him considers known and common standards of the manufacturing domain like OPC UA Unified Architecture as well as semantic models such as ISA95. This makes it possible to build an extensible, scalable information model that is capable of mapping process information as well as virtual and physical object representations of the production resources involved - from the machines to the end product. Through the use of such metamodeling, agents are able to independently learn and develop strategies that improve the overall process quality and performance indicators in a production plant.

In his work, Max Hoffmann has implemented his reference architecture into a testbed in which a virtual production plant implements autonomous planning and execution of a production process. This testbed was demonstrated at the Hannover Messe 2016 in combination with an interactive web application. The book by Mr. Max Hoffmann is a very good reading to understand the development and use of multi-agent systems combined with semantic standards in industrial applications.

Prof. Dr.-Ing. Tobias Meisen
Bergische Universität Wuppertal

Acknowledgement

This work was created as part of my employment at the Institute of Information Management in Mechanical Engineering (IMA) of the RWTH Aachen University throughout the past five years. During this time, I had the opportunity to work on exciting projects, talk to people of many different scientific backgrounds and learn a lot about various interesting topics.

First of all, I would like to thank my colleagues who accompanied me along this journey to my Ph.D. thesis. Great thanks are dedicated to Prof. Dr.-Ing. Tobias Meisen, whose scientific advise was extremely helpful during the process of my scientific development and in the course of this work. I would like to thank my working colleagues from the Production Technology research group at the IMA for many enlightening conversations and their feedback in terms of my research. I also thank all further colleagues at the Cybernetics Lab IMA/ZLW & IfU at the RWTH Aachen University for making this institute to such a special place.

Many thanks go to the various student coworkers, who spend days – and sometimes night shifts – in developing and finalizing programming code, applications or demonstration scenarios for the use-cases of this work. Among all of them, especially to mention at this point are Fabian Neumann, Fabian Scheidt and Stefan Angermüller for their extraordinary efforts during stressful working phases, e.g. ahead of the Hannover Fair 2016.

Very special thanks are dedicated to the students that pursued their master thesis under my supervision. Without you the present work would have never reached the scientific level and degree of maturity that it has today. Special thanks go the Stephan Groß for his work on interoperability in OPC Unified Architecture networks, Philipp Thomas for his developments on Multi-Agent Systems enabled by OPC UA, Jan Schlageter for targeting machine learning scenarios by means of the framework and Felix Wolf for his works on interoperability with high-level planning systems.

I thank the VDI/VDE-Gesellschaft Mess- und Automatisierungstechnik for enabling my participation in the expert group "Agentensysteme" (FA 5.15: agent systems) under the supervision of Prof. Dr.-Ing. Birgit Vogel-Heuser, in which I gained significant insights and inspirations for use-case ideas, especially in terms of the "myJoghurt" joint project. I thank the excellence cluster "Integrative Production Technology for High-Wage Countries" of the RWTH Aachen University and the DFG for their support during my thesis, especially for enabling the presentation of my demonstration scenario at the Hannover Fair 2016.

I am greatful for the supervision of my work by Prof. Dr. rer. nat. Sabina Jeschke from the Cybernetics Lab and Prof. Dr.-Ing. Birgit Vogel-Heuser from the Institute of Automation and Information Systems (AIS) at the Technische Universität München.

Last but not least, I thank my family and my friends for their support during all phases of my dissertation, especially my father Helmut for deep conversations and critical inspirations, and my mother Elisabeth as well as my brother Tim for all their encouraging words. You all play a big role in the completion of this work.

Max Hoffmann

Contents

List of Figures

List of Tables

Abbreviations

ACL Agent Communication Language

ADI Analyzer Device Integration

A&E Alarms & Events

AES Advanced Encryption Standard

AI Artificial Intelligence

AMQP Advanced Message Queuing Protocol

AMS Agent Management System

ANN Artificial Neural Network

API Application Programming Interface

AutoID Automatic Identification (*de: Automatische Identifikation und Datenerfassung*)

AutomationML Automation Markup Language

BACnet Building Automation and Control Networks

BDE Betriebsdatenerfassung (*engl.: production data acquisition*)

BOM bill of materials

CA coordination agent

CAD Computer-Aided Design

CAN Controller Area Network

CAQ Computer Aided Quality Management

CEP Complex Event Processor

CIM Computer Integrated Manufacturing

CISE Directorate for Computer & Information Science & Engineering

CNC Computer Numerical Control

CNP Contract Net Protocol

COM Component Object Model

CPPS Cyber-Physical Production Systems

CPS Cyber-Physical Systems

DA Data Access

DCOM Distributed Component Object Model

DCS Distributed Control System

DDL Device Description Language

DDS Data Distribution Service

DF Directory Facilitator

DHCP Dynamic Host Configuration Protocol

DIABOLO Distributed Architecture to Bolster Lifecycle Optimization

DL4J Deep Learning for Java

DML Dedicated Manufacturing Lines

DMS Distribution Management Systems

DPWS Device Profile for Web Services

DSS Decision Support System

DTACI Data Table Access Common Interface

EDDL Electronic Device Description Language

EMS Energy Management Systems

ERP Enterprise Resource Planning

ESB Enterprise Service Bus

EtherCAT Ethernet for Control Automation Technology

FDI Field Device Integration

FDT Field Device Technology

FIFO First In – First Out

FIP Factory Instrumentation Protocol

FIPA Foundation for Intelligent Physical Agents

GA Genetic Algorithm

GPIB General Purpose Interface Bus

GPIO general purpose input/output

GPRS General Packet Radio Service

GSM Global System for Mobile Communications

GUI Graphical User Interface

HDA Historical Data Access

HDC holonic architecture for device control

HMAC Keyed-Hash Message Authentication Code

HMI Human Machine Interface

HMS Holonic Manufacturing System

HTTP Hypertext Transfer Protocol

ICT Information and Communication Technologies

IEC International Electrotechnical Commission

IEEE Institute of Electrical and Electronics Engineers

IIC Industrial Internet Consortium

IIOP Internet Inter-Orb Protocol

IMS Intelligent Manufacturing Systems

IOM Inventory Operations Management

IoT Internet of Things

IP Internet Protocol

ISO International Organization for Standardization

IT Information Technology

JIT Just-in-Time

JPP Joghurt-Produktions-Protokoll

JST Japan Science and Technology Agency

KDD Knowledge Discovery in Databases

KPI Key Performance Indicator

LAN Local Area Network

LDS Local Disovery Service

LED light-emitting diode

LMS Least Means Squares

LSTM Long Short-Term Memory

M2M Machine-to-Machine

MAS Multi-Agent System

MAP Manufacturing Automation Protocol

C-MAPPS Modular Aero-Propulsion System Simulation

MDE Maschinendatenerfassung (*engl.: machine data acquisition*)

MES Manufacturing Execution System

MEXT Education, Culture, Sport, Science and Technology

ML Machine Learning

MOM manufacturing operations management

MQTT Message Queue Telemetry Transport

MRP Material Requirements Planning

MRP II Manufacturing Resource Planning

MSIP Ministry of Science, ICT and Future Planning

NSFC National Natural Science Foundation of China

NSERC Natural Sciences and Engineering Research Council of Canada

NSF National Science Foundation

OASIS Advancement of Structured Information Standards

OLE Object Linking and Embedding

OPC OLE for Process Control/Open Platform Communications/Openness Productivtiy And Collaboration

OPC UA OPC Unified Architecture

OS Operating System

OSI Open Systems Interconnection

OWL Web Ontology Language

PC Personal Computer

PHM Prognostic Health Management

PLC Programmable Logic Controller

PLM Product Lifecycle Management

PPC Production Planning and Control

PROFIBUS PROcess FIeld BUS

PROFINET Process Field Ethernet

Pub-Sub Publish Subscripe

QoS Quality of Service

REST Representational State Transfer

RFID Radio-Frequency Identification

RNN Recurrent Neural Network

RSA Rivest-Shamir-Adleman cryptosystem for public-key encryption

RTE Real-Time Ethernet

RTLS Real-Time Location Systems

RUL Remaining Useful Lifetime

SCADA Supervisory Control and Data Acquisition

SESA Semantically Enabled Service-oriented Architectures

SME Small and Medium-sized Enterprises

SOA Service-Oriented Architecture

SOAP Simple Object Access Protocol

SPoF Single Point of Failure

SSL Secure Sockets Layer

SVM Support Vector Machine

TCP Transmission Control Protocol

TLS Transport Layer Security

TSN Time-Sensitive Networking

UA-SC UA Secure Conversation

UDP User Datagram Protocol

URI Uniform Resource Identifier

URL Uniform Resource Locator

W3C World Wide Web Consortium

WAP Wireless Application Protocol

WLAN Wireless Local Area Network

WS Web Service

WS-SC Web Service Secure Conversation

WSMO Web Services Modeling Ontology

WSMX Web Service Execution Environment

XML Extensible Markup Language

XMPP Extensible Messaging and Presence Protocol

Glossary

Component The term *component* in the context of this work is utilized to denote *devices*, such as *sensors, actuators, mechatronic* or *infomechatronic devices*, as well as *machines, industrial controllers, software processes*, and *agents*. (see definition of Lastra (2006)).

Heterarchical Architecture / Control System A heterarchical control system is a flat structure composed of independent entities (agents). These agents typically represent resources and/or tasks. The allocation of tasks is usually performed by means of dynamic market mechanisms (Botti and Giret, 2008).

Holarchy A system of holons that can cooperate to achieve a goal or objective. The holarchy defines the basic rules for the cooperation of the holons and theirby limits their autonomy. (Botti and Giret, 2008).

Holon An autonomous and co-operative building block of a manufacturing system for transforming, transporting, storing and validating information and physical objects. The holon consists of an information processing part and often a physical processing part. A holon can be part of another holon. (van Brussel et al., 1998).

Intelligent Component The term *intelligent component* is also used to denote *components*, however by emphasizing not any kind of specialized production skills, but capabilities in terms of *reasoning, inference* and *learning* that enable *decision-making* (Lastra, 2006) based on some *contextual knowledge* and/or *perception* with regard to their surrounding environment.

Interoperability Ability of a system or a product to work with other systems or products without special effort on the part of the customer. Interoperability is made possible by the implementation of standards. (IEEE, 2016).

Kanban Kanban (literally signboard or billboard in Japanese) is a scheduling system for lean manufacturing and just-in-time manufacturing (Dictionary.com, 2011). Kanban is an inventory-control system to control the supply chain. Taiichi Ohno, an industrial engineer at Toyota, developed kanban to improve manufacturing efficiency. Kanban is one method to achieve Just-in-Time (JIT) (Öno and Bodek, 2008).

Knowledge The knowledge term that is referred to in this work focuses on the knowledge of smart entities in a manufacturing or automation environments. The knowledge of such agents consist of external input knowledge in terms of information models, ontologies and service requests as well as knowledge through an environmental perception.

Real-time The term real-time characterizes the operation of an information-technological system that is able to deliver intended results in a reliable way and within a time range that had been determined prior to the operation. Timely delivery of information might be according to a time grid or other time patterns, but does not have to be determined in a periodic way.
Important: In terms of this work, the real-time term is used in a "non-deterministic" way, focusing on *soft real-time* constraints. Thus, real-time is characterized as a short-term response in a few seconds rather than in the sense of deterministic *hard real-time*.

Abstract

This work presents a framework that enables the transition from grown manufacturing and automation systems that have evolved over long-term production life cycles to an interoperable factory of the future. This paradigm shift from tightly-coupled automation systems to loosely-coupled flexible information and communication infrastructures in manufacturing is performed by making use of object-oriented interoperability standards. The proposed framework facilitates the development of information models for manufacturing environments with the goal to enable autonomous decision-making based on real-time information from the field.

In terms of general demands for higher degrees of digitalization and the implementation of flexible manufacturing solutions within modern factories, major challenges to production planning and automation have to be faced in the near future. This work specifically focuses on the requirements of legacy and *running* systems with regard to technologies that are introduced by umbrella terms such as Industry 4.0 or Cyber-Physical Systems. As proposed in this work, the shift from current factories to digitized manufacturing environments is performed in two steps – the implementation of scalable information modeling and communication solutions into the factory environment and the establishment of autonomous systems.

The implementation of communication and interoperability solutions is performed by an introduction of interface standards that support both tightly-coupled architectures of legacy systems as well as service-oriented approaches for flexible information exchange mechanisms for loosely-coupled system architectures. With regards to these requirements the OPC Unified Architecture meta-modeling standard is utilized as it fulfills the demanded compatibility with deterministic bus systems and enables the integration into flexible networking environments at the same time. By making use of object-oriented information modeling capabilities of OPC UA it is possible to carry out digital representations of arbitrary factory environments.

This digital representation of a production site constitutes the starting point for the implementation of so-called smart agents into the factory environment. Using information modeling techniques, virtual representations of physical machines and devices are carried out by means of Cyber-Physical Systems. Smart software agents that coexist in both – the real and the digital – world are able to autonomously communicate, organize and optimize the production based on algorithms and dedicated software modules.

The realization of these smart manufacturing solutions is evaluated by means of industrial use-case that reflect the requirements and automation demands of today's production sites. A software demonstrator as well as validation scenarios are carried out to illustrate the applicability of the proposed solutions to real-world industrial processes.

Zusammenfassung

Das Ziel vorliegender Arbeit ist es, die Transition gewachsener Systeme in der industriellen Fertigung hin zu einer Fabrik der Zukunft zu ermöglichen. Der Paradigmenwechsel von eng gekoppelten Automatisierungssystemen zu flexibel verknüpften Informations- und Kommunikationsinfrastrukturen geschieht durch die Verwendung objekt-orientierter Standards zur Schaffung von Interoperabilität. Die vorgestellten Lösungen dienen dazu, eine informationstechnologische Abbildung und Modellierung von Produktionsumgebungen signifikant zu vereinfachen. Dies geschieht mit dem Ziel, Entscheidungsfindungsprozesse auf Basis von Echtzeit-Informationen aus der Produktion auf Basis autonomer Entitäten zu ermöglichen.

Durch zunehmende Forderungen nach stärkerer Digitalisierung und Flexibilisierung in der Produktion wachsen die Anforderungen an moderne Fabriken hinsichtlich der Implementierung zukunftsorientierter Lösungen für die Produktionsplanung und -steuerung. Diese Arbeit zielt insbesondere auf die Anforderungen an Altsysteme und laufende Produktionssysteme in der Fertigung ab, welche durch Technologien der Industrie 4.0 oder getrieben durch Begriffe wie cyberpyhsische Systeme modernisiert werden sollen. Für diesen Übergang präsentiert vorliegende Arbeit einen zweistufigen Prozess – zunächst erfolgt die Einführung von Methoden der skalierbaren Informationsmodellierung sowie Kommunikationslösungen, in einem zweiten Schritt die Etablierung autonomer Systeme in der Fabrikumgebung.

Die Implementierung von Kommunikations- und Interoperabilitätslösungen geschieht durch Einführung von Schnittstellentechnologien, die sowohl eng gekoppelte Architekturen als auch service-orientierte Ansätze flexibel gekoppelter Systeme für einen netzübergreifenden Informationsaustausch ermöglichen. Hinsichtlich dieser Anforderungen findet eine Anwendung des OPC UA Standards zur Metamodellierung statt, da dieser sowohl eine Kompatibilität mit deterministischen Bussystemen aufweist wie auch eine Integration in flexible Netzwerkumgebungen unterstützt. Durch die Nutzung der objektorientierten Modellierungsfähigkeiten von OPC UA ist es möglich, ein digitales Abbild komplexer Fabrikumgebungen zu realisieren.

Diese digitale Repräsentation einer Produktion dient als Ausgangspunkt für die Implementierung sogenannter smarter Agenten. Diese können unter Verwendung von Techniken der Informationsmodellierung dazu genutzt werden, virtuelle Repräsentationen physischer Maschinen sowie weiterer Komponenten aus der Produktion darzustellen und ermöglichen so die Etablierung cyberphysischer Systeme in den Produktionsverbund. Intelligente Agenten, die somit sowohl in der realen Welt wie auch im Cyberspace koexistieren, werden somit in die Lage versetzt, autonom zu kommunizieren sowie die Produktion auf Basis von Algorithmen und dezidierter Softwaremodule selbständig zu organisieren sowie zu optimieren.

Die Realisierung dieser verteilten, intelligenten Lösungen zur Produktionsoptimierung wird weiterhin durch eine Anwendung auf industrielle Use-Cases evaluiert, welche die Anforderungen realer Produktionsstätten widerspiegeln. Ein Software-Demonstrator sowie hierauf aufbauende Validierungsszenarien zeigen die Anwendbarkeit der vorgestellten Ansätze auf reale Anwendungsfälle in der Fertigung auf.

1 Introduction

1.1 Motivation

Recent technological advancements from customer sectors become an integral part of societal and economical reality. Data analytics, the rise of artificial intelligence and big data are major trends in all sectors of modern life. Especially industrial production is significantly affected by the changes that are implied through an application of these forecasting technologies.

In modern factories resources will be represented by intelligent entities that are organizing themselves, products that find their way through supply chains independently of management systems and machines that are able to reconfigure themselves in order to fit to the customer demands. Maintenance of production resources is anticipated in advance and unexpected failures of manufacturing resources can be compensated in a responsive manner.

Many application and significant economic potentials are opened up through the presence of these technologies. However, vital questions remain open: How can we provide all the information that is needed for these data analysis and forecasting use-cases? How is it possible to easily describe and aggregate vast amounts of information from various, heterogeneous sources in a way that condensed information is analyzable with an intelligent purpose? What is the definition of future interfaces that will enable communication between various parts and systems in modern factories? How can vertical interoperability from the sensors in the field up to the management levels of the enterprise be achieved?

The present work answers these questions by introducing a framework that combines native, interoperable interfaces with smart entities on the factory floor. These smart agents are able to represent every single entity, device, product, automation system and management level application present in a factory. A communication between these agents enables full data integration among various systems of different scope and scale. Rich information modeling techniques for the design of these agents ensure that production data is properly described by context information and can be analyzed by means of different purposes – for short-term and for long-term

© Springer Fachmedien Wiesbaden GmbH, part of Springer Nature 2019
M. Hoffmann, *Smart Agents for the Industry 4.0*,
https://doi.org/10.1007/978-3-658-27742-0_1

optimization of the production. The agents representing a vital part of the modern production will make use of algorithms to analyze this information in-process and to make decisions on-the-fly.

Data-driven approaches in service-oriented environments are able to make use of up-to-date information from the factory floor. In terms of this vision, smart entities combined with integrated (interface) technologies to obtain the needed information will finally enable a full exploitation of the potentials that are opened up by modern data analytics.

Besides this technology driven motivation of reaching towards a future oriented manufacturing, main drivers for the restructuring of modern production are also implied by competitive market conditions (section 1.1.1), ongoing trends in current manufacturing (section 1.1.2) and an increasing digitalization in modern factories (section 1.1.3). Possible challenges that have to be taken into account are addressed in the last part of this motivation (section 1.1.4).

1.1.1 Changing Reality – A Global Perspective

Due to the digital transformation and technological trends in almost all industrial areas, the world of production, manufacturing and industrial automation is facing major changes. Driven by distinctive merging of globalized markets, overarching availability of information and services as well as longer and more complex supply chains, modern enterprises need to regard themselves and their processes from a broader perspective.

In the light of data-driven technologies, machine learning and continuous data mining, the value of data and the retrieving of information becomes more important throughout all manufacturing domains. Techniques to obtained the required data are in need. These constantly growing challenges and requirement will likely bring current industrial state-of-the-art practices in factory and process automation to their limits (Leitão et al., 2016). Especially producing companies are facing a vast number of key challenges in terms of modernizing their current manufacturing and business processes regarding higher demanding and competitive market conditions without loosing their aims for a sustainable production.

To stay competitive with regard to these constantly changing market conditions, a completely new class of requirements in the field of production and automation technologies have to be met. One of the main requirements is to reach a high variety of customized products without risking the quality, timeliness and robustness of current manufacturing processes and the according products. For this purpose, the product life cycle in terms of the design, process implementation, production execution and control has to be renewed as efficiently and quickly as possible to reach the ability of reacting on constantly changing customer demands and market conditions. This reactiveness involves both, technological and organizational, e.g. information managerial advancements. These developments – especially from the technical communication's points of view – are not entirely realized yet. One core subject of this work is to address this lack of interoperability by establishing modern communication solutions enabling factories to react on changing conditions in a real-time manner.

1.1.2 Trends and Visions of Modern Production

The use of modern data analytics, reactivity in production and enhanced communication capabilities are also the major objectives of the so-called "fourth industrial revolution", representing the latest trends and technologies regarding disruptive changes within industrial reality. After the industry has already undergone three previous revolutionary transitions in terms of mechanization, electrification (Danielis et al., 2014) and automation/microelectronics, it is the aim of the ongoing proceedings to enable an Internet of Things (IoT) within the world of production that is characterized by intelligent, autonomous communication between smart entities representing all physical and information-technological systems of a production.

Single entities within current manufacturing sites already have these interconnectedness, e.g. enabled by wireless communication modules or open interfaces with access to their data. However, the IoT vision for production requires a more general and sophisticated approach that enables all present units, whether these are intelligent agents or relatively dumb sensors to be embedded within one overall context and communicate in a machine-readable way. This common *language* that is intended to serve as a bond between all systems, components and other interacting entities such as user access through Human Machine Interface (HMI) would bring the decisive added value towards a realization of such a web of things in production. One main objective of the architectural approaches presented in the course of this work is to define the establishment of such superordinate communication standard that is not only able to connect the entities of a manufacturing system, but also to enable virtual representations of hardware modules and to map their underlying information onto all levels and information systems of the company.

The goal of this work is to reach closer to the vision of the digital factory or a factory of the future by bringing together different strategies, planning layers, technical systems and current technologies. This is done by an interoperability approach that takes into account the information technological representations of all parts in a factory, i.e.:

- Semantic representation of data by means of context information to enable integration from heterogeneous sources into arbitrary information systems,

- interface technologies to incorporate systems that are characterized by information of different granularity and time scales,

- various information needs on different levels of the manufacturing.

The approach presented in this work intends to carry out a framework by means of scalable information representation that seamlessly works together with interface technologies and an intelligent usage of information in a real-time manner. Necessary steps on the way to these visions are the enabling of reconfigurability in production and the integration of smart devices into the core processes of manufacturing. The next section focuses on the integration of such smart devices from a manufacturing perspective.

1.1.3 Digitalization and the Introduction of Cyber-Physical Systems

One of the core ingredients that is connected to reconfigurable and adaptable production processes is the digitalization. Inspired by daily-life and omnipresent availability of information through digital media such as smartphones, calls for digitalized information in professional environments are constantly growing. The answer of the manufacturing industry to these aims consist in Cyber-Physical Systems (CPS), which function as an integration of computation and physical processes (Lee, 2008). In that sense, modern CPS are defined as "transformative technologies for managing interconnected systems between its physical assets and computational capabilities" (Lee et al., 2015; Baheti and Gill, 2011), hence facilitate a merging of process-related data and their digital representation. Accordingly, a "cyber-physical entity is one that integrates its hardware function with a cyber-representation for the physical part" (Leitão et al., 2016). This definition introduces the concept of an entity with a coexisting presence in both the real world as well as in a virtual cyber-space matching the requirements and boundary conditions of the real-world objects.

In that sense, the core of discussions around these cyber-physical systems lies in the establishment of a "digital twin" (Lee et al., 2013) that is able to serve as a digital representation of each production entity, such as machines, complex sensors or other components, and accordingly interoperate with other CPS, cloud platforms and/or higher information systems on behalf of these entities, exchanging information seamlessly through digital services. The presence of a digital twin as the true delineation of a cyber-physical system opens up various means to tackle the challenges and requirements imposed by the discussion around Industry 4.0 or the Industrial Internet as all process-related information such as data generated by various sources can interoperate within an IoT environment.

From a communication point-of-view, CPS are networked among each other with local digital communication systems but also incorporating connectivity to global networks. By interconnecting these local embedded systems with global networking environments, on the one hand numerous application of industrial use-cases and optimization opportunities occur (Danielis et al., 2014). On the other hand, the introduction of CPS poses the challenge in terms of merging features related to embedded systems like real-time requirements with the loosely character of web services and the Internet (acatech, 2011).

The aim of CPS and the digital twin is to generate a major impact on the interaction of all systems, components, products and human beings that are part of the manufacturing process or involved into the production in some other way. In order to characterize the broad field of applications that is affected by the establishment of CPS, the "5M system" describing the manufacturing domain in its entirety has been presented in Lee et al. (2013). Within the 5M approach, the five "M" stand for *Materials, Machines, Methods, Measurements* an *Modeling*.

Digital representations of physical entities in production are capable of addressing all of these manufacturing characteristics more effectively by integrating the 5M dimensions into one coherent context. Subsequently, through the integration of CPS into the lower levels of the production, all the technical process and quality related measures can be incorporated using either the physical entity itself in the real-world context or its digital twin for a representation and communication in the cyber-space. Machines, for example, can be regarded as physical entities that are on the one hand characterized by a certain process precision that can be optimized with regard to the physical process. On the other hand, machines also constitute resources that are characterized by means of their capabilities. The resource dimension of a machine is rather represented by its digital twin exchanging information with high-level planning systems such as Enterprise Resource Planning (ERP). This merging of physical properties and the *information dimension* is made possible through cyber-physical systems that are located within both dimensions, the real-world process and its digital counterpart. The interoperability approach demonstrated in this work integrates the CPS paradigm with scalable interface technologies eventually enabling a seamless communication throughout management systems and low-level devices represented by CPS.

Whereas the previous sections focused on the potentials from both, a technological and economical point-of-view, the next section concentrates on societal challenges that are implied by the digitalization and long-term implications connected to the vision of I 4.0.

1.1.4 Challenges of the Digitalization and the Industrial (R)evolution

Besides the various chances and opportunities that are created by an increasing digitalization of all technical areas, there might also occur certain risks and challenges that have to be overcome. During the latest "Digital Business Day" in Bonn, Germany, the CEO of the second largest IT enterprise in Germany, Software AG, mentioned that the "digital challenges are driving a wedge between established companies and their customers." (Billerbeck, 2016) He adds that the "digitalization is the most disruptive change that ever happened – due to its pace." As examples for the threat that is connected to an overall digitalization he especially points out companies like "Uber" and "AirBnB" who have the potentials of changing the market conditions drastically. As a suggestion to established enterprises he uses a rather radical comparison to the theory of evolution of Charles Darwin, when he mentioned towards current German enterprises: "Adapt yourself or become extinct!"

Long-Term Implications of the Digitized Factory

In that sense, the most important action to be performed by currently leading and well-established enterprises is to deeply understand the long-term implications that are connected to a digitized society and digital services. If the industry and especially sectors like mechanical engineering close their eyes to these challenges and do not adapt their products, services and also their decision making processes according to these novel challenges, they and their products are likely to become obsolete in the near future.

It is inevitable for manufacturers and product designers to adapt their life cycle to individualized products and manufacturing resources. As the producing industries and key players in information technology share common ground, a cultural shift in manufacturing towards a more agile decision making and adaptive behavior is needed. If this different culture between the developments in computer science or software and the product life cycles of industrial machines and automated processes will not be overcome, the technologies around Industry 4.0 are likely to end up in the "Through of Disillusionment", which is characterized as a low point following to a technology hype as defined by the "Gartner Hype Cycle" (Gartner, 2015). This effect is intensified by the formal requirements that are imposed by the industrial revolution in terms of standardization processes and the establishment of standardized metrics for the digital age of the machines. Successful and effective standardization in the context of Industry 4.0 might even be one of the biggest tasks during practical implementation. According to the study of acatech (Kagermann et al., 2013), a survey that was carried out among the members of the Platform Industry 4.0 (BITCOM, ZVEI and VDMA) came to the result that they regard standardization processes in general as the biggest challenge within the new era of manufacturing. The critical issue in this context is to facilitate the decision making in terms of new standards and procedure models in manufacturing without loosing the robustness, solidity and reliability of yet proven approaches.

Soft and Hard Challenges on the Way to Industry 4.0

With regard to the discussed risks and opportunities of the cyber revolution that is taking place, the evaluation and implementation of Industry 4.0 has to be performed from a broader perspective, taking into account not only technical and information-technological challenges, but also issues that affect the environmental and human implications of the industrial revolution. As the life cycle of automated production systems is changing in an increasing pace, the pressure and requirements on manufacturing enterprises increases significantly (Lüder, 2014). Driven by the speed of technological innovations within the information technology domain, which advances almost on a daily basis, the development and enhancement of automated production systems need to be adapted in a disruptive way to reflect the pace of

technological developments that are connected to seamless interoperability
and overarching availability of information.

In that sense, a transition from the manufacturing reality – as it is
right now – and a factory of the future has to be designed carefully and
deliberately considering grown manufacturing and automation systems. The
technological "upgrade" of these so-called "brown-field" plants into future-
oriented applications using innovative and up-to-date technologies (e.g.
modern interfaces as well as information integration, storage and processing
of data) has to be performed smoothly and in accordance with the people,
who need to adapt these systems into their everyday work reality.

One key requirement of this work is take into account especially existing production facilities and their according emerged ecosystems that are
characterized by a composition of technical resources and human beings.
All innovative solutions – which may be of revolutionary or evolutionary
nature – and technological changes have to be introduced by considering
the technological-environmental boundary conditions of each plant, taking
into account domain-specific requirements and worker-related conditions.

1.2 Research Goals

**The goal of the current work is to derive a new concept for the
representation of smart agents on the shop floor through semantic
interface technologies. Current approaches only target high-level
planning algorithms or optimization on the shop floor. This work
brings together both concepts – by integrating up-to-date information from the shop floor seamlessly into high-level planning systems. The OPC UA standard is used to derive a next-generation
interoperability concept by representing agent communication in
a semantic context for integrating low-level automation data into
the loop of smart manufacturing and reconfiguration.**

One of the most essential steps is to create information transparency within
the factory, i.e. collect all data that is being produced by various sources,
aggregate as well as transform this data into useful information and finally
integrate these valuable pieces of information in an accessible way. The
result of this step is to reach an information readiness that enables a target-
oriented usage of production data in a holistic and real-time manner for
adaptivity and reconfigurability of manufacturing processes.

The next step in pursuing the path to a real-time factory is to integrate not only the components of the manufacturing environment, but also to enable interoperability between production systems and higher information systems of the management layer of an enterprise by transferring interoperability approaches from an information technological point of view into the purposes of a technical environment like a factory. By doing so, the essential issue is to integrate loosely-coupled systems – as they are present within networking topologies – and tightly-coupled systems that can be found in most automation systems such as field bus applications and which are characterized by rigid structures and control loops. Bringing together these two worlds is an integral – and might also be the most critical – step towards an interoperability between technical infrastructures and management systems.

A natural interoperability between technical systems and abstract information management environments delivers the basis for a descriptive context that enables CPS to be embedded into and used within a manufacturing environment. CPS, which are also part of both worlds – the physical reality and the cyber-space – rely precisely on the interoperability mentioned above. The descriptive context of CPS hereby delivers the according framework to annotate production data with information that is needed to interpret the data on higher levels of planning/management systems and to integrate the meta-data into CPS that are situated on the edge of the physical processes and represent all information that is generated in the field. This approach finally enables a coupling of smart and reactive manufacturing environments with intelligent high-level planning and optimization systems through the CPS paradigm and opens up opportunities for an intelligent production.

Based on a functioning and well-defined approach for embedding CPS into arbitrary manufacturing environments, the next step consists in using the computation capabilities of cyber-physical systems to establish smart entities, which are also referred to as smart agents, into the field level of a production. For this purpose, an additional requirement of these arising smart CPS is to enable the needed connectivity of CPS with one another and at the same time with external systems located on higher levels of production organization systems. The aim of creating internet-inspired (e.g. IoT) architectures in production is only reachable, if seamless communication and information exchange with systems of different data representation or granularity can be guaranteed at all times. This demand is pursued by integrating the modeling, communication and high-level interoperability of smart agents by means of established interface and communication standards that ensure a consistent representation of information and the communicating entities.

Finally, the last step of pursuing the way to a reconfigurable factory is to make use of the enabled scenarios by means of learning approaches that are opened up through the scalability of agent-based system architectures. This can be done by applying machine learning scenarios into the context of smart agents that able to make use of the integrated information from the factory floor.

With regard to the research goals described above, the following research questions have been shaped to sum up the fundamentals and goals of this works' approach:

Research Question 1 – Which are the basic model definitions that are needed to design cyber-physical production systems by using agents that are able to integrate tightly-coupled and loosely-coupled systems on the shop floor?

Taking into consideration present manufacturing environments it is inevitable to annotate low-level production data with context information in order to integrate this data into a higher context and to use the integrated information in a real-time manner. The answer to the first research question delivers a *descriptive context* that enables data exchange of field level information by means of a digital twin that is located on the shop floor and interoperates seamlessly with loosely-coupled top-level systems.

Research Question 2 – What is the next-generation interoperability standard that incorporates low-level communication and high-level system logic? Due to the indispensable presence of established management systems for resource and execution planning in most factories, ways to integrate information from the shop floor in a generic way are in need. This includes not only networking devices, but also legacy systems based on traditional automation solutions. The *central research question* of this work concentrates on interoperability concepts for modern production systems by means of the next-generation interoperability standard OPC UA and its service-oriented communication approaches. Answering this question enables a holistic usage of up-to-date information from smart agents on the shop floor in a real-time manner.

Research Question 3 – Which is the most effective design approach to a service-oriented architecture in production that is capable of integrating smart agents with high-level management systems?

The answer to this research question finally delivers the reference architecture that is capable of addressing the challenges and obstacles of current manufacturing aiming towards a future-oriented production by means of service-oriented approaches and intelligent interfaces all at once. For this purpose, a specialized interface agent is introduced that is able to incorporate ERP systems with smart agents in the field.

Research Question 4 – What are the contributions of the derived architecture to enable a reconfigurable production based on autonomous, learning agents? The last research question concentrates on the validation of the derived approach by means of a machine learning scenario that enables anticipatory manufacturing in terms of a predictive maintenance use-case. The realization of this scenario into a scalable multi-agent system architecture demonstrates the potentials of the derived approach.

The goal of this work opens up a perspective for additional research questions that will be enabled by the proposed framework and information-technological concept. Further research activities in the field can be carried out on the basis of the scalability approach chosen for the model-driven design of domain-specific information models as shown in terms of this work. One of the most important questions for such future activities is:

Open Research Question – By means of which concepts is it possible to automatically design proper information-technological representations for production infrastructures and semantics by means of domain-specific information model extensions?

The last question – representing the open research question and opening up further ideas for future works – addresses the scalability of the proposed architectural approach. Taking into account the advanced interface standards that are used for the communication within the proposed architecture, further methods and proceedings for enhancing the scalability of this approach in terms of additional application domains or detailed product specifications are in focus. In terms of this research questions, ontology-based approaches will be further utilized to enable such integrability of standardized information models into the factory and in the representation of smart entities on the shop floor.

The outcomes of answering these research questions deliver the key elements to form a reference model for the design and implementation of smart, self-organizing and self-adapting manufacturing systems that are characterized by intelligent entities communicating with each other and with all environmental systems in a seamless and real-time manner.

1.3 Structure of the Work

The structure of the work follows an iterative approach, in which the challenges of current manufacturing systems in terms of a digital transformation are pointed out prior to an examination of the technological requirements and implementations needed to reach a digitized factory. The overall goal of the work is to motivate and to build up the basis for a next level interoperability framework taking into account current technical restrictions and limitations as well as future demands. According to this target, the parts of the work are presented as follows (see Figure 1.1).

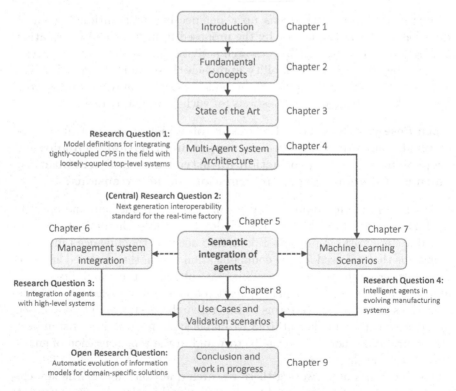

Figure 1.1: Structure of the work and allocation of research questions

Chapter 2 introduces insights and the most important fundamental challenges of current factories. The underlying problems are pointed out against the background of automation systems that have been emerged over decades of industrial manufacturing. This special character of the production domain in contrast to other industrial sectors poses special problems and restrictions to a modernization of current factories. The entire industrial production sector was and will always be primarily focused on high efficiency. Thus, currently running and comparatively well established systems are difficult to update using new technological approaches, as any idle period within industrial manufacturing is connected to high breakdown costs. Chapter 2 deals with this sort of special preconditions of current automation systems and characterizes the difficulties imposed by an emerged information and communication technology architecture in modern factories.

The state of the art section describes the organizational and technological standards regarding today's industrial manufacturing practice. The first parts of chapter 3 focuses more on general standards, paradigms and framework attempts without going into detail about their technical feasibility or implementation. Further parts of the section cover the underlying technologies of the described superordinate architectural approaches and point out details about their technical feasibility. The state of the art section closes with the examination of technologies that represent the spirit of future industrial standards and unconventional approaches to deal with both, the limitations and the potentials of current automation systems.

Chapter 4 picks up the ideas introduced in the state of the art section and accordingly describes the derivation of a general architectural approach that makes use of ideas connected to smart entities and CPS in the manufacturing domain. The presented approaches take into account limitations of the current situations in manufacturing companies and start with a requirement analysis in terms of the desired framework. Subsequently to this analysis, the derivation of an architecture based on a Multi-Agent System (MAS) of cooperating intelligent entities seamlessly communicating with the devices of automation systems on the shop floor is presented. The framework architecture is carried out by means of agent-based approaches as well ideas inspired by holonic manufacturing systems and independently of obstacles and issues of their realization. The chapter closes with a critical discussion regarding the limitations imposed by current implementations.

Chapter 5 provides means to realize an architecture for smart entities autonomously planning and executing a manufacturing process. This chapter contains the main contribution of this work and derives architectural design patterns to map the representation and communication of smart entities on an object-oriented meta modeling approach. This approach allows for an integration of smart embedded systems with various other systems of a manufacturing enterprise. The models derived in terms of this concept represent the aforementioned descriptive context for CPS within a flexible, interoperable and reactive infrastructure.

Chapter 6 demonstrates the applicability of the derived framework architecture in terms of incorporating management applications in terms of a enterprise resource planning system. It will be outlined precisely, how the scalable information modeling approach of the proposed agent model contributes to a generic interface towards high-level systems and how information from arbitrary enterprise application can be considered in the manufacturing process.

Chapter 7 extends the derived CPS infrastructure in terms of learning agents. These learning capabilities are demonstrated by means of machine learning scenario that focuses on predictive maintenance in the manufacturing domain. The applicability of the concept is shown and validated by means of a quantitative price function that enables negotiation techniques between intelligent agents by taking into account up-to-date information from the manufacturing system.

Chapter 8 represents the use-case and validation chapter of this work. In order to show the applicability of the derived approaches, a demonstration testbed had been carried out in the course of this thesis that represents a network of embedded devices autonomously organizing production processes through an integration of different technologies seamlessly working together. The information modeling approach is validated by the implementation of a multiple domain-specific model extensions using agent-based approaches through OPC UA.. The learning capabilities of the agents are shown by means of an evolving negotiation mechanism within the network of cooperating agents.

Chapter 9 introduces future research topics that are opened up by the proposed framework. Main fields of action for future investigation will consist in the a generic information model generation, e.g. in terms of ontology matching or other techniques that enable a dynamic extension of the domain or application-specific topics addressed in this work.

2 Problem Description and Fundamental Concepts

This chapter deals with the fundamental concepts of current manufacturing systems and problems need to be overcome on the way to a factory of the future. The first step to approach these challenges consists in a characterization of paradigm shifts that are needed to address the information transparency demands of a factory that is able to reconfigure itself by means of up-to-date information from the field. Fundamental concepts of modern industrial manufacturing deliver a brief description of industrial manufacturing prior to a basic characterization of emerged manufacturing systems in current factories. The last section of this chapter accordingly introduces the concept of bringing together system architectures of different scope and of different time-scales for reaching a real-time interoperable factory.

2.1 Strategical decision-making in a factory of the future

The paradigm shift from high-quality, but uniform products to product portfolios with high variety and customer-orientation imposes agile decision processes in terms of organization, production planning, implementation and reorganization of manufacturing processes not only within the planning phase, but also during the process itself. Design and process decisions that have to be made in early planning phases of the production organization will have a major impact on both the management levels of manufacturing enterprises and at the same time on process-related layers of the manufacturing execution planning, i.e. cycle times, optimization cycles and process implementations have to be designed in an adaptive and changeable way. During these decision processes different concrete targets, long term goals and time scales have to be taken into account in order to reach a holistic optimization of the processes from a strategical and technological point of view. The merging of these yet separated systems – located on the management levels and the operational layers of a production – will be addressed in this work

© Springer Fachmedien Wiesbaden GmbH, part of Springer Nature 2019
M. Hoffmann, *Smart Agents for the Industry 4.0*,
https://doi.org/10.1007/978-3-658-27742-0_2

by means of an architecture that incorporates communication technologies suitable for all subsystems of an enterprise ecosystem including various functional layers and components. Earlier works of the author already focused on these sort of issues, however with an emphasize on measurement data (Meisen et al., 2013).

Decisions that need to be performed from a *management perspective* of the enterprises according to new demands regarding flexibility have an impact on both, mid-term and long-term results. These interdependencies and influences from a management perspective are likely to undergo major changes, as production planning based on long-term customer relations is no longer feasible in a competitive market environment and will inevitably switch towards agile decision-making due to varying demands. Hence, in future-oriented enterprises, management decisions influencing short-term goals of the production will not be uncommon in the daily practice.

From a *process point of view* though, mid-term planning processes as well as short-term scheduling will play an even more decisive role for the process organization in future manufacturing, as changeability requirements will finally strive for in-process adaptations in a real-time manner. By taking into consideration these different long-term, mid-term and short-term goals that affect both, management decisions as well as decisions during tactical planning, the complexity of according Decision Support Systems (DSS) increases significantly as more determining variables, boundary conditions and volatile parameters have to be considered. The architecture and interoperability framework that will be presented within this research work, will contribute to the merging of these different decision objectives by bringing together the information needed at management levels and the information that is capable of realizing a seamless adaption of shop floor activities and execution planning. This shift of current manufacturing sites towards service interactions on different levels is currently the major show stopper, but at the same time the biggest chance for improvements in the future, based on an establishment of service-oriented paradigms in the factory environment.

To achieve the demanded product variety and process adaptability from a *technological perspective*, manufacturing systems are required to become more flexible, robust and configurable by supporting agile and adaptive responses to changing conditions through a dynamic reconfiguration of their ongoing processes (Leitão et al., 2012). However, if manufacturing systems are designed to be more flexible, reconfigurable and accordingly reach a higher level of distribution, their complexity also increases decisively

(Trentesaux, 2009). This complexity augmentation does not only affect the hardware-near shop floor applications, it also complicates the aggregation of information and accordingly their integration into high-level information systems. It is one of the main obstacles on the way to a "factory of the future" to deal with this complexity in a structured and organized manner. Despite the discussion among the fourth industrial revolution (acatech, 2011), the definition of the "factory of the future" has already been coined nearly 30 years ago by Welber (1987):

> "The factory of the future will resemble a very large scale intelligent machine that operates with a highly integrated and well-organized base of knowledge. It must be flexible with regard to changes in demand, technology, economic conditions, and competitive pressures. Above all, it must accurately measure and interpret both what the customer wants and what it can make."

Despite the fact that the goal of this definition did not loose its actuality, the technological requirements, resources and tools to reach this vision have been emerged substantially since this time. The key challenges of the characterization above, however, have not been completely solved yet, as bringing together these technological achievements into one coherent infrastructure is still the biggest task of the upcoming developments in terms of Industry 4.0 (Kagermann et al., 2013). Merging technological approaches and currently isolated solutions for dedicated tasks has yet to be achieved in order to provide means for enabling the production of the future, which is characterized by complex systems and sub-systems working together in a seamlessly cooperative manner.

Reconfigurability in production

Reconfigurability is one central demand of modern manufacturing processes to reach the above-mentioned requirements in terms of flexibility, adaptiveness and responsiveness to customer demands. Reconfigurability as an enabler for future-oriented factories is one of the major characteristics of changeability in general, which is defined as "the characteristics to economically accomplish early and foresighted adjustment of the factory structure and processes on all levels including the enterprise, in response to change impulses" (ElMaraghy et al., 2012; Wiendahl et al., 2007). In this context, changeability is described as an "umbrella framework that encompasses many paradigms such as agility, adaptability, flexibility and [especially] re-configurability, which are themselves enabler for change" (ElMaraghy et al., 2012). According to Barbosa (2015), the term Reconfigurable Manufacturing Systems (RMS) is defined as a concept that suggests the rapid change in a factory's structure using changes in hardware and/or software to adjust the production capacity and functionality in a quickly manner. In the context of manufacturing systems, i.e. in terms of reconfigurable machinery, flexibility capabilities are characterized as follows (Abele et al., 2006): (i) The choice of a concrete machine that is capable of performing a certain operation from a given set of capabilities; (ii) variation of different production sequences for finishing a given product and the (iii) variety of different capabilities a machine can achieve with short changeover time.

These adaptation capabilities increase the complexity of a production system significantly as such reconfigurability is only reachable by means of a vast number of data sources for determining the current system state, and at the same time enabling actuators to inject these adaptations back into the process. The complexity imposed by bringing together an arbitrary number of technological tools and paradigms demands for new ways of organizing industrial manufacturing processes and their according information flows for automation purposes. In the light of these challenges ideas of decentralizing decision making come to mind, as the centralized management and organization of these highly complex structures is hard to achieve.

Challenges of reconfigurability in the light of Industry 4.0

Current efforts in terms of tackling these challenges regarding increased complexity imply a need for formal and structured approaches that support the decision making both in the planning phase and during execution of the manufacturing. With regard to these structural approaches, terms like "Industry 4.0" (Kagermann et al., 2013), which was coined by the German Government, or the "Industrial Internet" (Lin and Miller, 2015) dominate ongoing discussions in terms of reaching a new level of technology in manufacturing as an enabler for reconfigurability and adaptability of production processes. The concept of the "fourth industrial revolution" discusses approaches and applications for the factories of the future and their goals. However, many of the proposed use-cases, e.g. in Bauernhansl et al. (2014), or applications are presented rather on a conceptual level than giving concrete recommendations for the implementation and realization of such scenarios. A rather comprehensive example for the realization of an

Industry 4.0 use-case in terms of an open-source framework can be found in one of the author's contribution (Hoffmann et al., 2016d).

Research activities and challenges for upgrading current factories

The visions as described above are made up in countless research papers, presentations and projects, however a vast number of the proposed approaches, frameworks or reference architectures do not address the explicit challenges that hinder the establishment of these frameworks, which are *legacy systems* that need to be incorporated in all future-oriented factory automation solutions. Other obstacles are imposed by the strong penetration of current automation systems with proprietary shop floor solutions, such as outdated bus systems and deprecated communication protocols. Solution approaches that do not consider these historically grown ecosystems and boundary conditions are not suitable for tackling the real challenges. As a matter of fact, recommendations that contain the prerequisite of replacing all present machines and sensors of a current manufacturing environment with "new ones that are Industry 4.0 capable" are useless for small and medium-sized enterprises, which have neither the financial capabilities nor the time to renew their entire system configurations and requirements.

The current work gives deeper and more detailed insights about the occurring problems and restrictions that have to be taken into account while upgrading current manufacturing sites according to the paradigms of the technological requirements behind an omnipresent call for "Industry 4.0". The challenges that have been described above both concerning the decision making from a management perspective and with regard to the adaptability and flexibility of manufacturing systems lead to the same underlying problem, which is caused by a lack of relevant, up-to-date information from the production. It is information availability in particular that enables most of the use-cases implied by Industry 4.0, may it be for customized products, varying customer demands or energy saving potentials in a factory.

The Duality of Information Needs and Information Overload

When looking more closely on the issue of this information shortage in manufacturing enterprises, there is an interesting duality which comes to mind. As a matter of fact, the lack of information describes only one part of the problem. Another issue, which is also part of the same problem, consists in the information overload that is also a major topic modern enterprises have to deal with. Caused by an increasing number of physical and digital data sources being part of the production, problems of processing and handling the exploding amount of generated information poses an equally severe challenge. An answer to both problems can only be given by an information management that structures the vast amount of data according to a consistent data or information model. An efficient and well-structured modeling approach is, on the one hand, able to deal with a high number of information sources by processing and integrating them in a defined way. On the other hand, an information modeling approach also provides means to structure raw-data according to an appropriate description with meta-data in order to convert the data into useful information.

The Introduction of Smart Data to Deal with High-Volume Information

This useful information does not only serve the basis for profound decisions in the planning phase and execution of the production. It also facilitates the processing of huge data streams from the manufacturing site by constantly transforming raw (or "Big") data into "Smart Data" that contains not only information about the data point, but also about the context information that is necessary for a meaningful and automated interpretation of that data. One desired goal to deal with high amounts of raw data from the shop floor would be an automatic assignment of information from the lower levels of the factory. An automated annotation of production information with context information such as meta-data would reach both, machine-readable and interpretable information for an autonomous process optimization as well as a data basis understandable by humans. Information that is provided in a human-readable form helps operational managers and executives beans means of Key Performance Indicators (KPI), which provide a broad overview about relevant business process networks. According to these advantages implied by a profound and consistent information modeling of data sources in an enterprise, the first step that is necessary to enable a future-oriented production in current manufacturing sites, consists in an establishment of well-defined ways for data description and information consolidation for all processes that affect the planning and optimization of the production in an enterprise (Hoffmann et al., 2014).

The availability of up-to-date information on all levels of the production execution and organization enables the final goal of reaching reconfigurable manufacturing systems. In opposition to the more technically motivated definition of reconfigurable manufacturing system, the characterization according to ElMaraghy (2005) defines an RMS as a "[. . .] manufacturing systems paradigm that aims at achieving cost-effective and rapid system changes, as needed and when needed, by incorporating principles of modularity, integrability, flexibility, scalability, convertibility, and diagnosability." In order to reach these dynamic reconfigurability goals and to provide up-to-date information to react in a flexible and optimized way on constantly changing conditions, real-time information has to flow through all different layers from the shop floor up to management and business process level. In modern, distributed environments, in which information has to be exchanged through an institution-wide or even cross-institutional service landscape (Colombo and Karnouskos, 2009), the goal of the forecasting manufacturing enterprise is to reach a flexible, loosely-coupled service-orientation towards the real-time enterprise and hence a factory of the future.

Research Goals Regarding Reconfigurability in Modern Production

The goal of this research work in terms of reconfigurability demands is to enable flexibility of production processes by providing an approach that enables efficient information management in factories by means of scalable, generic information modeling concepts to enable a holistic utilization of real-time information on all levels of the production in an enterprise and for the intelligent behavior of smart entities on the shop floor and on the execution layers of the manufacturing. For this purpose, appropriate interface and information modeling concepts are identified that enable not only an aggregation of production information for monitoring and mid-term optimization, but also to realize an intelligent usage of the information in real-time by means of autonomous, smart entities. The emerging framework intends to facilitate the development steps of digitalization and information-technological networking of current manufacturing sites to an information and communication infrastructure that complies with the latest standards of factory automation and modern approaches towards a real-time factory. In that sense the goal of this work is not to propagate an approach by means of *replacing all production resources with new ones*, but especially to integrate legacy systems that rely on different cycle times than new systems of Information & Communication Technology (ICT).

2.2 Fundamental Concepts of Industrial Manufacturing

Despite the state of the art in manufacturing and automation technology, there are some paradigms and characteristic fundamental problems existing on a higher level than the problems that are motivated by technological issues. Thus, the key challenges mentioned in this chapter are rather of historical or socio-economical than of technological nature, as they have been evolved within automation systems throughout the last decades. The manifestations of this evolutionary process can be regarded in terms of rigid control hierarchies, proprietary protocols and steady states in the context of process planning, implementation and adaption, which are all characterized by well-established methodologies and rigorous proceedings and are not likely to change in the near future.

The reasons behind these established methodologies are versatile. On the one hand, production processes and thus machines and tools within industrial manufacturing environments are characterized by rather long cycle times. Hence, extensive investments into new hardware or the establishment of new communication standards in automation environments is always connected to an economical risk as long depreciations and financial commitments need to be considered. On the other hand, there are also major show stoppers in connection with safety, security and reliability issues and their according solutions that have been emerging throughout the last decades. During this era, several standards, norms and guideline had been developed by standardization organization that work in close contact to industrial key players. Hence, due to the development of these standards in connection to industrial adopters the applicability of the derived guidelines is proven by real-use cases. However, the emerging of those standards are mostly also connected to some sort of market-political interests of the big players in the automation industry.

The fundamental problems of process stability and reproducibility with regard to production processes are fairly solved by current standards and guidelines, as the processes derived by such standards are characterized by rigid and reliable behavior. Thus, developments away from these standardized and formal development policies are connected to high risks, in case new methods, communication standards or machine interfaces do not meet previously assured requirements of a stable process. This risk is the key issue that is responsible for the relatively slow adaption rate of Industry 4.0 solutions especially by Small and Medium-sized Enterprises (SME) (Andersch et al., 2015). These companies will only adapt new trends in automation technology, if their current system architectures and manufacturing configuration can be basically maintained, while innovative technologies such as recommended by the Industry 4.0 only add some sort of added value in comparison to the actual state.

Hence, the fundamental motivation of maintaining such ecosystems in industrial practice is understandable, due to the reasons mentioned above. However, focusing on conventional methods or proceedings in planning and manufacturing execution is mostly at the expense of flexibility and adaptability of the manufacturing processes. Taking into account these competing trends and issues, one key requirement demanded by such trends as the fourth industrial revolution or the Industrial Internet is to enable a subtle transition from current practice in production automation towards modern approaches that are characterized by an overall adaptability, reconfigurability and flexibility of the manufacturing processes enabled by advanced communication standards and information integration concepts.

Technical challenges resulting from that aim of upgrading current manufacturing sites and their outdated underlying technologies with modern approaches is to perform such transition without risking the stability and functionality of currently running systems. As already mentioned in section 1.1, the ability of incorporating emerged industrial ecosystems into any kind of advanced architectural approach is one of the main assumptions this research work relies on. Thus, in the course of all described developments, legacy systems and their according restrictions, limitations and boundary conditions are always considered as one cornerstone of the framework respectively the proposed reference architecture.

In order to examine the limitations and boundary conditions of emerged ecosystems, the fundamental objectives of legacy systems in manufacturing environments are described in the following sections. Therefore, current production respectively automation systems and their fundamental technologies are characterized first, before the duality of bringing together rigid and flexible control/automation systems as the central prerequisite for establishing new approaches in industrial manufacturing is described subsequently.

2.3 Emerged Manufacturing Systems in Industrial Production

In this section, the general boundary conditions and prerequisites of current factories and their industrial automation environment are drawn together. This compilation focuses more on a higher classification of these systems than on technical details of the underlying technologies, which are describes in detail within the state of the art chapter. After the general characterization that is stated in this chapter, the limitation and drawbacks of these systems with regard to interoperability in production are highlighted.

2.3.1 Static Control Systems in Current Factories

Current factories are optimized for the production of high-quality, uniform products. Many industries, therefore, are making use of a so-called Dedicated Manufacturing Lines (DML): "DML, or transfer lines, are based on inexpensive fixed automation and produce a company's core products or parts at high volume [. . .]. Each dedicated kind is typically designed to produce a single part (i.e. the line is rigid) at high production rate achieved by the operation of tools simultaneously in machining stations." (Koren et al., 1999) The authors further mention that with regard to this kind of production coordination the product costs correlate to a high demand for uniform products, hence high customer demands lead to relatively low overall production costs. Accordingly, the most economical and efficient state of such manufacturing lines is achieved at full capacity production (Koren et al., 1999). However, not only with regard to the high competition worldwide, but also due to varying conditions and specific customer demands, this manufacturing methodologies might be doomed to fail.

One expression of such varying conditions are unpredictable changes of product designs and connected specifications for their engineering process. This often leads to the necessity of changing process plans that are usually connected to costly and time consuming changes to machinery configurations, manufacturing plans and automation programs. An agile and responsive change of the according process planning is a vital requirement for modern factories. However, during the conduction of current methods for planning or re-planning, new process plans from the scratch are derived nearly every time a change takes place (ElMaraghy et al., 2012). Hence, DML, as defined above are mostly characterized by high delays and unpredictable costs in terms of change or scalability requirements.

Due to the limited scalability and changeability of dedicated manufacturing lines, process optimization and further improvement within these processes are mostly performed by enhancing dedicated production steps or by adjusting the overall configuration of the production. Optimization and design of optimized processes are derived for each of the process steps separately and are eventually implemented into the manufacturing entity at machine level. This implementation usually comprises of a fixed program that is loaded into control units that are directly attached to the machines and other automated components of the manufacturing. Subsequently, all the logic of the production process is concentrated on the lowest level of the manufacturing organization.

This lowest level of the production is defined as field level, and represents the layer that is closest to the process (John and Jasperneite, 2011). The process-related data generated by the machines and other manufacturing entities on the field level constitutes the highest amount and frequency of information. At the same time, the data from the field level is also characterized by the lowest density of information as it usually comprises of *raw machine or sensor data*. Some applications on higher levels of the manufacturing enterprise might depend on the full amount of data from the field level, e.g. Supervisory Control and Data Acquisition (SCADA) systems or Manufacturing Execution System (MES), other systems might depend on a different granularity of information, e.g. a higher density, i.e. the information is only useful, if it is further aggregated in order to form useful indicators, e.g. for proceeding usage in ERP or Product Lifecycle Management (PLM) systems. When looking at these different sort of systems and their variety of target objectives, it becomes clear that one of the main obstacles on the way to reconfigurability and flexible planning approaches lies in the ability to exchange information between these systems in a seamless and autonomous way, without loosing the meaning of information and at the same time providing the data in a machine-readable form. Over time, the different systems described above have been emerged to some kind of structure that shows interdependencies between the systems and information flows between different layers in the form of a pyramid.

2.3.2 The Automation Pyramid and Hierarchical Production Organization

The introduction of a hierarchical representation for field level networks occurred in close connection to first attempts of making data available across all functional layers of a manufacturing enterprise. The resulting automation pyramid (see Figure 2.1) comprised the first multilevel network model defined for the purpose and the scope for Computer Integrated Manufacturing (CIM), an initiative that was initiated to "cope with the anticipated complexity of data in a horizontally and vertically integrated communication environment" (Sauter, 2010). The original establishment of the pyramid resulted in the definition of five levels that characterize the different layers of the factory (Sauter, 2007).

The hierarchical organization of the systems that are involved in production planning and execution is formally described by the automation pyramid, which constitutes the traditional way to manage manufacturing from top-level (ERP, PLM) to the bottom (field level). The shape of the pyramid indicates the increasing density of information from the field level to the top of the pyramid, where the information is available in a condensed form (Vogel-Heuser et al., 2009). The majority of data sources, however, is located on the bottom, in the shop floor respectively on the field level.

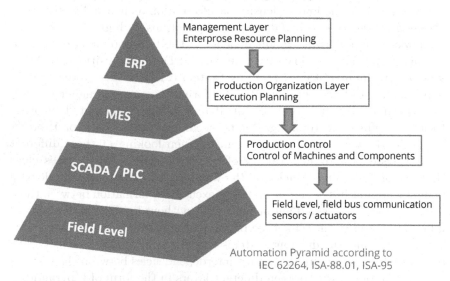

Figure 2.1: The automation pyramid of manufacturing and control

The automation pyramid is characterized by a control flow that is induced at the top and flows to the bottom. The information flow though is intended to go from bottom to the top.

In the top level information systems, namely ERP and PLM systems, the enterprise or plant-wide planning and configuration of the production takes place. The ERP system accordingly manages the resources of a plant such as machines, conveyor belts, transports units and material all along with human resources. PLM systems take into account the entire life-cycle of a product in order to optimize product configurations and the economic success when designing, constructing and manufacturing of a product. Thus, in both cases, these planning systems very much rely on a high density of various information from miscellaneous sources. Most of these sources are located at the bottom, where the machines and sensors of the production process generate the actual information which is linked to the physical processes taking place in the manufacturing.

Directly above the lowest layer of production – the field level – the Distributed Control System (DCS) and SCADA systems as well as production control units such as Programmable Logic Controller (PLC) are located. This layer is directly, i.e. physically connected to the shop floor and thus benefits from an immediate access to raw data from the field and is able to perform direct interactions with devices such as actuators or machines. A Programmable Logic Controller hereby carries a logical (static) program that is used for the automation of electromechanical processes, such as control of devices on the shop floor. PLC usually contain several digital and analog inputs and outputs and are able to deal with noisy signals and other process-related characteristics of the manufacturing. Whereas PLC control a single production process or assembly line, distributed control systems generally control a number of distributed components such as distributed PLC located in several production lines throughout the plant.

Supervisory Control and Data Acquisition systems are also characterized by an immediate interconnection to the field, but include rather diagnostic than control functions. More specifically, SCADA represent remote monitoring and control systems that are connected to the shop floor via a fixed communication channel, usually by means of bus systems. The "data acquisition" part of such systems represents the functionality of using status information of the remote equipment for visualization or persistence purposes (Cyber Security Dictionary, 2012). In practical applications the functionality of a SCADA system might include abilities for automatic monitoring of e.g. critical values such as temperature signals as well as functions for displaying information for human beings, e.g. maintenance workers in the field.

Besides the automation pyramid, other standards also describe the cooperation between the systems on different levels of the production hierarchy, especially from a product-oriented perspective. A standard that attempts to divide the complexity of manufacturing systems into four layers is the ANSI/ISA-95 standard according to IEC 62264 (International Electrotechnical Commission, 2013) (see Figure 2.2). This standard especially focuses on these layers in terms of defining how and where decisions for the manufacturing are made. The layers that are covered by the model are the control (Level 1 and 2), operations (Level 3) and business (Level 4) (Barbosa, 2015).

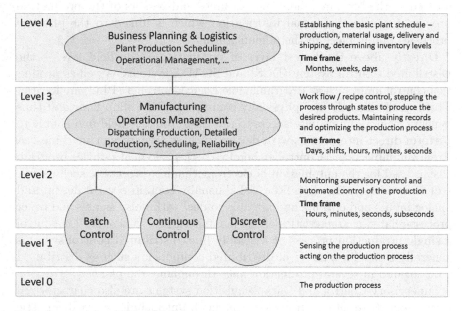

Figure 2.2: The ANSI/ISA-95 standard

On the lower Levels 1 and 2 the control of actual manufacturing equipment is performed with the objective to manufacture products. These control actions are realized using PLC, various resources, Computer Numerical Control (CNC) and SCADA. Level 3 represents the complement of the MES layer and determines a number of complex activities, such as detailed scheduling, quality management and maintenance. These activities are performed to prepare, monitor and complete the production that is executed in the lower levels. Level 4 is the highest level, representing the ERP layer and is being used for strategical decisions regarding the enterprise.

Both of the basic architectural models shown in terms of the automation pyramid and the ANSI/ISA-95 standard attempt to facilitate the manufacturing structure by means of a hierarchical model. In both models, an important component of the manufacturing organization is represented by intermediate layer between the shop floor and the management systems. For the ANSI/ISA-95 standard, this intermediate layer is the Manufacturing Operations Management, the automation pyramid summarizes the functions of work flow organization and control under MES functionalities.

The MES serves as the medium layer between the high-level enterprise applications at the top and the low-level hardware-near components in the shop floor and above. According to McClellan (2001), the "MES system bridges the gap between the planning system and the controlling system using on-line information to manage the current application of manufacturing resources: people equipment and inventory." Hence, the objective of an MES in a modern factory is to function as an interface between planning systems and the manufacturing execution by transforming the information from high-level systems into concrete tasks to be executed on the field. McClellan (1997) states further that MES have been "build around the manufacturing" (McClellan, 2001) and are therefore "more than a planning tool like ERP [... more] an on-line extension of the planning system with an emphasis on execution and carrying out a plan". The tasks of MES comprise the following main objectives, which are directly connected to manufacturing execution (McClellan, 2001), such as *making products, measuring parts, turning machines on and off, moving inventory to work stations, setting and reading measurement controls, changing order priorities, scheduling and rescheduling equipment, changing order priorities, assigning and reassigning personnel and inventory.*

This list of functionalities shows that MES are somehow located in *both worlds*, the planning and management world as well as in the shop floor / field level of a production. This means, MES are already able to perform operations demanded by a future-oriented factory. The major role of MES in this context is to service a detailed planning to manufacturing execution and at the same time deliver real-time information from the field into higher systems, i.e. serving transparency. Kletti (2007) mentions with regard to MES that "a necessary condition of process transparency is the ability to map the company's values stream in real time, without [laborious] acquisition processes." Kletti (2007) adds that this capability related to MES is "beyond the [capabilities] of the dominant ERP systems."

In other words: The key to holistic production optimization taking into account resource planning, value streams, product life-cycles and machine/ sensor data is to utilize and exploit this information altogether in real-time. The availability of information allows for a "holistic view of production and services equipment and facilities detailed planning, status detection, quality, performance analysis, material tracking, etc." (Kletti, 2007)

The central challenge arising from this vision is, how to enable modern middleware systems like MES to "understand" the various information that is generated in the field and within the different systems that are responsible for the other components of the holistic value stream mapping of the production. One major paradigm that complicates the realization of this vision is the duality of loosely and tightly-coupled systems, which will be described in section 2.4. In order to get an idea about the different nature of these tightly and loosely coupled systems, the general concept of tight coupling in manufacturing is described first.

2.3.3 Tight Coupling in terms of Conventional Automation Infrastructures

According to the automation pyramid depicted in Figure 2.1 the lowest level of the production is represented by field devices performing the actual manufacturing steps and serving as data sources in terms of measurement variables, etc. These systems on the shop floor rely on hard real-time capabilities, as these originally relatively "dumb" devices needed exact and detailed instruction about which step to perform instantly and in a determinable timely manner. These hard real-time capabilities are the major reason for the existence of such systems nowadays. The easiest way to ensure these real-time requirements is to interconnect the according device using fixed, wired connection, in which every data package is determined in direction as well as in terms of the payload that needs to be transferred. These rigid topologies, which are mostly characterized by point-to-point connections, define the tight coupling.

Due to their determined behavior and predictable time of data delivery, tightly-coupled systems are able to meet real-time capabilities through a predefined system behavior. The fixed wiring of all devices by means of serial bus connections is what these systems make so reliable and function so well for non-changing use cases and system configuration. This durability, however, is inevitably connected to a lack of flexibility and adaptability of the ongoing processes, as unknown use cases cannot be predicted and calculated with the same accuracy as known ones. Eventually, every adaptation of the system or process configuration is connected to unpredictable efforts as all time constraints and system behavior needs to be set-up each time something in the system environment has been changed.

2.4 The Duality of Loosely and Tightly-Coupled Systems

As described in section 2.3.2, the key to success in optimizing the production in a holistic way is to integrate information from all levels of the automation pyramid in real-time. However, this intention is difficult to realize as the "two worlds" of a manufacturing system are subjects to a completely different scope and pursue fundamentally distinctive aims, which are on the one hand complete information transparency and information management capabilities for high-level systems and on the other hand, hard real-time capabilities in the field level.

There were early ideas addressing attempts of incorporating field level networks with the rest of the automation hierarchy as shaped by the automation pyramid, e.g. in Wood (1987). However, as a matter of fact, these attempts have never really been considered by the industrial key players developing those standards. Thus, the structured integration of the field level into higher systems was "never the primary intention" (Sauter, 2010).

Thus, as discussed above, it is in fact one of the major issues of bringing together these systems that the "different levels of the pyramid are controlled by mutually largely incompatible networking concepts" (Sauter, 2010). These different topologies are characterized by fieldbus systems in the lower levels and mostly Ethernet- ans IP-based Local Area Network (LAN) environments. The difficulties that are connected to the integration issues between these two networking concepts were always and are still one of the main promoters of using Ethernet on the field level (Sauter, 2010). However, even if the two networking worlds of shop floor automation systems and office floor networking systems can be brought together through some Ethernet approach, this attempt only diminishes one part of the problem. The mere interconnection of the different layers of the pyramid through one common networking environment does not solve the interoperability issue on its own, as it only delivers the fundamental basis for data exchange (Sauter, 2010) between these systems, but does not specify how this information exchange, i.e. based on which semantics, is performed in a practical application.

Far more important than barely enabling a suitable networking environment would be to use a common protocol basis, i.e. the Internet Protocol (IP). This "non-Ethernet-specific property [...] actually alleviates data exchange across the [vertical] levels of the automation pyramid and makes the pyramid structure flatter [...]" (Sauter, 2010). Eventually, this leads to a structure, in which the layers of the pyramid interoperate in a non-hierarchical way (see Figure 2.3). In the course of this transformation, information between different systems – and *not* layers – of an enterprise are exchanged based on common interfaces. This approach would finally enable interoperability between these systems by means of generic services.

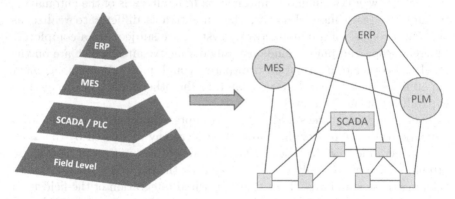

Figure 2.3: Transformation of the automation pyramid to a flexible system

3 State of the Art

The first part of the state of the art section is focused on general trends and requirements on the way to a factory of the future. Current strategies and paradigms to optimize automated production processes are pointed out first, before further parts of this section will go into detail regarding technical realization by means of concrete technology stacks.

Based on general introduction in terms of the requirements regarding a hierarchical organization of the manufacturing processes in a producing company, further sections of this chapter will concentrate on their technical realization within the different hierarchical layers of the enterprise structure. The technologies involved are primarily focused on reaching vertical integration throughout the factory and enterprise layers in order to reach the overall goals and an optimized automated production flow.

Section 3.1 gives a brief introduction into current attempts of integrating technologies made possible by an omnipresent digitalization in terms of general reference architectures for the Industry 4.0 or similar calls for action.

Section 3.2 gives an introduction of current enterprises' aims in terms of optimization and continuous improvements strategies for automated industrial production technologies. The limitations of these scenarios lead to requirements in terms of more sophisticated interoperability and communication solutions covering the entire value chain of a producing enterprise.

Section 3.3 will focus on the general requirements of modern automation systems in terms of interoperability as a general enabler for a digitalization of devices and their networking capabilities. Current interoperability approaches are examined and their limitations regarding the needs of future-oriented factories are pointed out.

Section 3.4 examines the current state of the art for automated production systems in terms of technical solutions located on the shop floor. This includes a general introduction into established field bus protocols including their way of modeling and automating manufacturing processes. The limitations of current approaches especially in terms of scalability, continuous information modeling and integrative communication lead to a need for more sophisticated concepts and communication solutions.

© Springer Fachmedien Wiesbaden GmbH, part of Springer Nature 2019
M. Hoffmann, *Smart Agents for the Industry 4.0*,
https://doi.org/10.1007/978-3-658-27742-0_3

Section 3.5 goes into detail regarding modern networking strategies throughout the entire company taking into account the requirements of an automated production in the field as well as the requirements of an information exchange with high-level systems of an enterprise. The next steps that go beyond traditional interoperability solutions based on sole protocols or syntactical standards is to provide standardized means of addressing various resources and services in a complex manufacturing systems by means of a Service-Oriented Architecture (SOA). Solutions based on web services and similar technologies in terms of a SOA aim at interconnecting various enterprise layers, e.g. through service buses, enabling vertical interoperability between different information (management) systems.

Section 3.6 concentrates on autonomous systems in manufacturing. By interconnecting various devices any sort of other entities through scalable service interfaces, intelligent behavior based on complex choreographies or other interconnected process flows can be realized. One possible and very widely adopted way of transferring the idea of intelligent behavior into the manufacturing domain by means of service-oriented approaches are MAS. Thus, the main part of this section focuses on current developments related to intelligent software agents.

Section 3.7 focuses on the evolving of intelligent software agents in terms of machine learning and similar evolutionary strategies. A basic characterization of machine learning as well as their applications in the manufacturing domain are pointed out.

In order to make use of intelligence and learning approaches that are made possible through the introduction of MAS into the production domain, new ways of modeling the manufacturing environment and the according agents have to be carried out. Thus, in section 3.8 possible ways of describing the manufacturing environment in accordance with the integration of smart CPS are pointed out. The information modeling of such smart entities enabling a *digital twin* sort of representation of the manufacturing resources will be described as follows.

Section 3.9 is dedicated to an advancement of the information modeling ideas described in the previous section. In order to deal with more complex communication needs between different participating entities of the manufacturing process, sophisticated semantic standards for horizontal and vertical information exchange mechanisms have to be deployed. Semantic interface standards that go beyond information exchange in terms of specific protocols are the answer to these communication demands. OPC Unified Architecture (OPC UA) as a holistic communication standard for the industrial production offers common interfaces along with a metamodel specification for in depth information modeling and will be presented in detail during this last part of the state of the art section.

3.1 Global Strategies towards Future Manufacturing

With regard to the digitalization in most industrial sectors, general attempts to target a general restructuring of the major challenges have been shaped in the recent past. On the one hand, these attempts are accompanied by reference architectures approaches (section 3.1.1) that aim at providing software stacks for a common standardization. On the other hands, important research programs around the world try to target the global challenges by high-volume research programs, of whom the most important are mentioned in section 3.1.2

3.1.1 Reference Architectures for an Industry 4.0

As a characteristic example for such reference architectures, the *RAMI 4.0* model (Koschnick, 2015) has been shaped by several major institutions involved in the high-tech strategy of the German government with regard to the requirements of a general digitalization strategy in industrial production. Another attempt to generalize the building blocks of future-oriented factories in characterized by the *Industrial Internet Consortium (IIC)* with the introduction of their reference framework – the *Industrial Internet of Things (IIoT)* (Lin and Miller, 2015).

Despite different origins of these attempts for general technology stacks, the main components and aims of these reference architectures target similar approaches. These approaches and possible solutions regarding their technical realization are outlined in order to characterize the limitations of the technology stacks currently used.

3.1.2 Important Research Programs targeting the Global Challenges

The key requirements that were mentioned in the fundamental concepts and their prospective solutions are distinctively inspired by the ideas of CPS. In the line of these research areas, the key requirements for future-oriented CPS by means of interoperable approaches have been investigated by various research projects in terms of different initiatives, programs and roadmaps dealing with similar problems of the manufacturing organization and CPS-inspired means to solve them. According to Leitão et al. (2016) the most important of these research programs have been originated in terms of the *Horizon 2020* program that characterizes the CPS paradigm as a key pillar of the "Factories of the Future". Other research activities worth to mention are present among nearly all continents of the world:

- The *Industrie 4.0* research program of the German government, which especially focuses on CPS in terms of the "vision whereby intelligent objects collect, store and process all their own relevant information throughout their life-cycle, and systems composed by these objects benefit from all the information and knowledge that is actively generated" (Leitão et al., 2016).

- The Swedish program *Made in Sweden 2030* along other similar programs in Spain that also focus on the ideas implied by the Industry 4.0.

- The Austrian government also decided in 2015 to spend about 250 million Euro by means of pursuing projects in terms of bringing together the CPS ideas and applications of the Industry 4.0.

- The United States concentrates on the research on CPS under the umbrella of the Directorate for Computer & Information Science & Engineering (CISE), the National Science Foundation (NSF), the U.S. government as well as major enterprises such as AT & T, Cisco Systems, General Electric, IBM and Intel under the name *Industrial Internet Consortium (IIC)* (Lin and Miller, 2015).

- Similar programs are present in Canada, where Natural Sciences and Engineering Research Council of Canada (NSERC) support several research activities in terms of CPS.

- The Ministry of Education, Culture, Sport, Science and Technology (MEXT) in Japan sponsors research activities regarding CPS through the Japan Science and Technology Agency (JST).

- China announced independent research activities in terms of the initiative *Made in China 2025* to support the development of intelligent and green manufacturing while focusing especially on quality, adaptation and integration with the Internet (Leitão et al., 2016). The according R & D projects are managed by the National Natural Science Foundation of China (NSFC).

- Finally, the Korean Ministry of Science, ICT and Future Planning (MSIP) supports research activities that are dedicated to the CPS field.

3.2 Current Approaches for Production Optimization

Actual approaches for reaching an optimized planning and execution of automated manufacturing processes can be distinguished into several categories, focusing on different granularity levels of the production and varying in terms the approaches used to reach global optimization goals. The most important of these approaches are examined in the following.

Certain strategies primarily focus on the *organization of the production*, i.e. by reaching an optimized allocation of automated processes. These kind of static optimization approaches rely on fixed production plans and are usually not distinctively changed during an actual production process (section 3.2.1).

Other approaches for optimizing automated processes concentrate on the actual usage of information from the shop floor, e.g. by taking into account information that is collected through data acquisition facilities such as sensors, control units and similar operation devices. A collection or aggregation of these data sets is generally utilized in terms of *midterm and long-term optimization strategies* and will be examined in section 3.2.2.

Other more sophisticated, future-oriented approaches concentrate on an automatic usage of production information from the field. Processes, which are optimized in terms of these strategies, are summarized under the term of *self-optimizing (manufacturing) systems* and will be further examined in section 3.2.3.

3.2.1 Organizational Optimization of the Production Process

Some high-level approaches for optimizing automated production processes are mostly focused on the organization and management of these processes in terms of general requirements and characteristics of the production flow. Widely applied proceedings of these strategies relate strongly to the domain of discrete manufacturing aiming at the ultimate goal of *continuous-flow manufacturing* (Leone and Rahn, 2002). This sort of production organization is – in contrast to batch production – primarily focused on a *just-in-time* and *kanban* production approach. Central requirements regarding this sort of improvement strategy are related to an ultimate integration of all elements of a production system. Continuous-flow manufacturing sites are characterized by four major goals that are intended to characterize an optimally *balanced*, lean production:

- Prevention of waste as the highest goal of the manufacturing flow. In this context, the definition of waste is intended not only in terms of material waste, but also regarding waste of time and human resources.

- Lowest-possible cost by choosing the optimal set of parameters for all subsequent processes and investments is another way of optimizing the production flow.

- The aim of *on-time* production strongly correlates to the prevention of waste and underlines the interest of lean production approaches in solving problems as quickly as possible.

- Finally, the *defect-free production* underlines a very important issue as the total costs or total waste of the production often highly correlates to necessary rework demands due to defects. Thus, a successful prevention of defects often results in a significant reduction of waste.

Approaches for reaching these goals are often pursued by means of dynamic batch sizing and flow-time reduction, summarized under the term of *repetitive flow manufacturing* (Jacobs and Bragg, 1988). The basic assumption behind these dynamic batch sizing ideas is that a continuous flow – unlike for example chemical processes – is difficult to achieve when dealing with discrete products such as cars. Higher batch sizes usually lead to longer waiting time until a batch of discrete products can be transferred to the next production step. The goal of the *single-piece flow* paradigm is to avoid this kind of waiting time waste by reducing batch sizes drastically (Liker, 2004). On the other hand, reducing batch sizes is also connected to an increase in organizational, logistical and/or management overhead.

For many years, the strategical efforts of the management in producing companies were focused on the above described optimization strategies. The aim of this strategic behavior is usually connected explicitly to a long decision horizon. In terms of these long term strategies, an incorporation of entrepreneurial goals with the production strategy of the enterprise is desired. According to Hill (2000) and Günther and Tempelmeier (2012) the integration of market strategical goals with production strategies is performed in five steps (Metz, 2014):

1. Determination of enterprise objectives regarding general key figures like revenue, growth and profitability. The strategical outline of the

formulation of these targets is performed by taking into account the enterprise vision and the general planning horizon.

2. The second step is connected to carrying out a market strategy. This strategy defines the classification of target markets, product portfolios and planning of intended target customers and sales volumes.

3. The next step consists in the determination of product-related policies and has a direct impact on the competitive factors of the market strategy. These policies specifies target figures with regard to product pricing, product design, product quality and services connected to the market launch of the product portfolio.

4. Based on the strategy and competition related product policies determined in the last step the technical realization and the concrete production planning is targeted in the next step. The technical planning consists of a specification regarding manufacturing methods, basic resource planning, flow of material and level of automation.

5. During the last step, the infrastructure of the intended manufacturing system is carried out in terms of resource allocation, plant layout, quality management life cycle and human resource planning. The infrastructural design is strongly connected to the product policies and the general production planning determined within the previous steps.

These steps can be regarded as a recurrent life cycle of products from a manufacturing enterprise. Especially the last two steps are likely to change during the evolution of a product portfolio. Although this general classification of the necessary steps in terms of the ramp up management of products did not lose its actuality, the product life cycles especially in terms of adapting a product or certain manufacturing steps has decreased rapidly. Thus, despite the fact that the entrepreneurial goals of producing companies have not been changing significantly in the last couple of years (step 1 to 3), the indispensability of accelerating the manufacturing and product life cycles become obvious (steps 4 and 5). In order to meet this requirements, the feedback loops between the process related low-level system and the managerial decision-making need to be evolved to the next level. Current approaches of enabling close feedback loops between the shop floor and high-level systems in order to reach profound decision-making by the managing instances is described in the next section.

3.2.2 Long-term Optimization of Production Goals using Low-Level Data

Shop floor interoperability approaches as described in the previous section aim at serving horizontal interoperability among different entities in the field level. This sort of interoperability is characterized as a necessary requirement for the goal of flexible and reactive automation processes. However, in order to meet sufficient prerequisites of a reconfigurable manufacturing system, further measures have to be taken into account enabling a sustainable adaptability of manufacturing systems. Especially in terms of a constantly increasing number of shop floor entities to be considered in system optimization, the development of decentralized, thus distributed intelligent control systems is inevitable.

In terms of these intelligent control systems, it is further required that the intelligent processes are more closely developed in direct interaction with the physical system. This fact is especially relevant in industrial control systems that "are required to control widely distributed devices in an environment that is prone to disruptions" (Brennan, 2007). Despite the research that has been already performed in terms of high-level production optimization, i.e. for scheduling, planning, supply chain management and enterprise integration applications, a seamless integration of these optimization techniques with up-to-date information from the shop floor that relies on hard real-time requirements has not been sufficiently focused yet.

However, this close interaction between planning systems and low-level information integration is needed as "current industrial automation systems are expected to operate in inherently complex and unpredictable environments that consist of a wide range of resources distributed across large areas." (Brennan, 2007) Most of these systems though, are being carried out by means standardized, large-scale computer systems that are not connected closely to the processes on the shop floor. The results are system configurations with computer hardware platforms that support monolithic control applications. Thus, when setting up system configurations based on these monolithic systems, it usually takes months until an automated production site shifts into the operational phase (Brennan, 2007) and any changes are time-consuming as well as difficult to realize without disturbing a running system.

3.2.3 Autonomous Optimization Strategies in terms of Self-Optimization

Due to the lack of generic interoperability between shop floor and the high-level systems another approach is to design control mechanisms by means of autonomously optimization systems. One approach that goes beyond conventional solutions of closed control loops is summarized under the terms of *self-optimization*. The self-optimization of production processes hereby characterizes a control methodology that opposes the previously covered planning-oriented optimization approaches (Brecher, 2011).

Planning-oriented approaches usually attempt to provide a detailed description of the entire production process by characterizing all field level approaches as detailed as possible. This kind of specification is mostly performed by taking into account vast number of parameters, production properties and product policies including their interdependencies. As a result, a set of rules determines the behavioral reaction of the production to any possible system state. The complexity of this approach, however, usually requires a restriction of the parameters sets taken into account to the ones that are determining for the actual process. The determining parameters that are being selected during this modeling process intend to describe the system behavior not precisely, but with reasonable accuracy. This approach aims at two target goals: (i) The first one is to provide a procedure that avoids unexpected uncertainties during the process by enabling rational, model-based decisions; (ii) on the other hand, this approach also attempts to significantly reduce the complexity of the system characterization by simplifying the modeling where justifiable (Auerbach et al., 2011).

In manufacturing processes, though, the identification of relevant parameters that are in focus of the production optimization are not easy to determine (Knowles and Nakayama, 2008). Thus, most of the optimization models that are currently employed are specified for a determinable number of stationary production situations rather than for arbitrary systems states. Due to the high variability of modern manufacturing scenarios these static model approaches do not provide the required flexibility. Accordingly, possible changes in system configurations and inner structures due to varying boundary conditions and manufacturing prerequisites are mostly associated to tremendous efforts towards an adaptation of process models (Cisek et al., 2002). Accordingly, in terms of these restrictions, ramp-up processes or the introduction of new product components or similar product policy adaptations are connected to a reorganization of the entire model.

Accordingly, flexible ways for adapting manufacturing control models are in need, which autonomously determine unexpected situations and actively anticipate forms of adapting to these new conditions. In terms of self-optimization approaches, these scenarios are mostly performed by means of adapting closed control loops, e.g. as shown in Permin et al. (2015)

Tremendous research efforts have been carried out in the field of self-optimization for production systems, e.g. in terms of the *SFB 614 – Self-Optimizing Concepts and Structures in Mechanical Engineering* (Gausemeier et al., 2008a,b). A explicit coverage of these attempts would go beyond the scope of this work. However, as one major aim of this thesis focuses on intelligent behavior on the shop floor of a manufacturing process, approaches that go beyond this general idea of adaptive control loops will be discussed in section 3.6.

3.3 Interoperability Approaches for Manufacturing Systems

Interoperability between heterogeneous systems always poses major challenges to developers and system operators, no matter whether the targeted application is located in the consumer segment or within business applications. Especially in terms of manufacturing enterprises, whose communication is based on severely heterogeneous systems using different architectures and design principles, interoperability is rather difficult to reach and to maintain.

In the consumer segment, solutions for interoperability already exist. However, mentioning transmission media such as cable, Wireless Local Area Network (WLAN), Global System for Mobile Communications (GSM) or General Packet Radio Service (GPRS) omits the fact that all of these – internationally accepted – standards only define solutions for the transport layer of the communication. An appropriate platform for enabling interoperability within an information and communication technology architecture relies several additional requirements to ensure their functionality in a heterogeneous environment (Hoppe, 2014):

- Diverse services for accessing information sources such as sensors, actuators or data storages and information sinks such as databases. Other services enable connectivity to HMI and enable identification of users and an access through Application Programming Interface (API),

- Appropriate semantic description of services and interfaces – in order to enable autonomous communication and generic interoperability not only among human users, but also between machines, embedded devices and information management systems, adequate context description for the services in use need to be provided. These descriptions need to precisely declare the purpose and usage of the available services. This formal description of services and interfaces for accessing these services are indispensable for offering functionalities and interfaces in a machine-readable form,

- Standardized ways to discover and recognize devices as well as an unambiguous description of their functionality. The identification of devices and components has to be independent of the number of networked entities (scalability) and independent from the transport protocol that is being used for accessing the device.

There are already several approaches that have been developed either by research institutions or by industrial key players that aim for an "upgrade" of current manufacturing systems in order to comply with Industry 4.0. According to Schumacher (2015), a Director of a leading MES provider, the transition from the current state of the art within industrial automation to Industry 4.0 is divided into four stages, of whom each needs to be fulfilled in order to comply with the technological demands of an Industry 4.0. These stages comprise:

Stage 0 – Status quo The manufacturing environment is characterized by rather instable production processes and a low degree of automation and organization. From the IT perspective, only ERP systems and some other isolated solutions are available for automatic configuration and technical design of the production process.

Stage 1 The first stage is achieved by a stable production process through automated data acquisition and integration using MES and/or systems like Betriebsdatenerfassung (*engl.: production data acquisition*) (BDE), Maschinendatenerfassung (*engl.: machine data acquisition*) (MDE) and Computer Aided Quality Management (CAQ). This configuration already provides a high level of automatic integration, however the degree of organization is still low.

Stage 2 Decentralized MES that provide capabilities of reactive and detailed planning. Consequently, these systems are to perform a production segmentation as well as a continuous flow of production, e.g. based on First In – First Out (FIFO) and Kanban organization. The "flow of production" is a basic principle of the famously known "Lean Manufacturing" system developed for the production at "Toyota". The term was essentially coined by Öno and Bodek (2008) through the definition of a Just-in-Time (JIT) model for optimizing the intra-logistics of a production.

Stage 3 The third stage is characterized by full vertical and horizontal integration according to established interoperability and MES integration standards (e.g. VDI 5600 (VDI, 2012) or ISA-95 (ISA, 2000)). The goal of this level of Industry 4.0 compliance is a full supply chain integration including suppliers and customers.

Stage 4 The final stage for Industry 4.0 readiness is achieved, when all production resources are linked to each other through dynamic represent- ations by CPS. The integrative data management is performed by MES. The processes are further characterized by a distinct decentralization of all decision processes for the manufacturing organization. Thus, through full Industry 4.0 compliance, flexible production systems are finally achieved.

The last step of the described evolutionary chain constitutes the readiness for Industry 4.0 solutions, in which all information from different levels of the production comes together and is used by autonomous smart entities to optimize the manufacturing in a cooperative manner. The interoperability needed to perform such behavior in a real production site might be limited to certain constraints that are present in current manufacturing systems. In order to define these limits, the general aims and target objectives of the desired interoperability in production are described in the following.

From a broader perspective, interoperability describes the ability to exploit the entire added value of useful information that is generated based on production data. The aim of this interoperability is especially to make this information accessible on all levels of a manufacturing enterprise. The availability of information that is generated in the field and within shop floor applications for other systems, e.g. management or planning tools, can only be realized, if interoperability in the described way can be successfully established between these systems. In this context, interoperability "means the ability of autonomous and independent systems to cooperate without any help by external units" (Epple, 2011). This sort of cooperation requires that these different systems *understand* each other, i.e. exchange information in a machine-readable way.

The problems that occur when analyzing the information from different levels of a manufacturing enterprise, especially the data that is generated on lower levels of the automation pyramid, consists in the different characteristics and granularity of information compared to other levels. In order to extract useful information from these systems and to use it in higher levels of the production planning and organization, the *raw data* from field level has to be enriched by context information to provide all additional information that is necessary to understand or classify the according pieces of information in terms of their origin or with regard to the circumstances, in which this data has been recorded.

3.4 Technical Solutions for Automated Production Control

This section examines the technical background of automated production systems and describes the historical developments that have emerged from the underlying technologies of different information exchange mechanisms in industrial automation. The investigation starts with a closer look at fieldbus protocols, which are not only the oldest technology used for industrial communication (Felser, 2002), but also still represent the largest share in industrial applications of the automation practice (Dietrich and Sauter, 2000). Secondly, the Industrial Ethernet as the next step towards an enhanced, interoperable communication in industrial networks, incorporating the aforementioned fieldbus systems, are examined. Industrial Ethernet hereby represents the industrial counterpart to standard Ethernet, which is by no doubts the most dominant network technology used in Local Area Networks (LAN) in the office world and in terms of most Internet-based applications (Spurgeon and Zimmerman, 2014).

After the presentation of these basic and well-established communication solutions for the factory shop floor some more advanced technological approaches are pointed out in the following, mostly aiming at tackling the real-time issues that cannot be solved with conventional methods, especially for large scale, distributed systems with many different, widespread information consumers. The Publish-Subscribe (Pub/Sub) mechanisms represent a light-weight, scalable way of exchanging semantic-rich information with arbitrary, distributed components by means of various networking topologies and configurations.

Finally, some more advanced Industrial Ethernet applications are examined in terms of real-time capabilities, especially with regard to standardization attempts in terms of Time-Sensitive Networking (TSN), a standardization attempt by the IEEE that concentrates solely on the issue of realizing real-time capabilities in industrial network environments. These next-generation Industrial Ethernet standards aim at completely replacing conventional fieldbus systems and former Industrial Ethernet standards within industrial automation by incorporating the hard real-time functionalities of traditional field devices through a distinctive reservation of bandwidth in the network. Due to its networking capabilities, this next-level Industrial Ethernet is also capable of integrating the flexible behavior of Pub/Sub mechanisms by using their information modeling and information exchange approaches.

3.4.1 Fieldbus Protocols

Fieldbus systems are part of an industrial computer networking protocol family and are standardized in DIN EN 61158-1 (2015). Fieldbuses are primarily used for real-time control of distributed systems. The development of fieldbus systems is based on the paradigm that complex industrial automated systems require a distributed control system that organizes, manages and controls an organized hierarchy of controlled systems. According to Sauter (2010), fieldbus systems comprise the first milestone in the evolution of field-level networking systems, followed by Industrial Ethernet and wireless approaches for the networking within industrial automation systems. Outlining the historical relevance of fieldbus systems, these "field-level communication systems have been an essential part of automation for [more than] a quarter of a century [...]. More than that, they have made automation what it is today." (Sauter, 2010) The term *fieldbus systems* originally refers to the fields of process engineering, e.g. especially for chemical plants (Sauter, 2005). The definition that is provided by the International Electrotechnical Commission (IEC) is rather imprecise as it primarily underlines the fieldbus standard as some sort of programmatic declaration than on the technical characterization of these systems: "A fieldbus is a digital, serial, multidrop data bus for communication with industrial control and instrumentation devices such as – but not limited to – transducers, actuators, and local controllers" (DIN EN 61158-1, 2015).

More detailed information about the technological background and the history of fieldbus protocols can be found in Appendix A.2. Despite the fact that these additional information are not of particular interest for this work, the interested reader might refer to this background as a motivation of modern interface standards that are discussed in later sections.

3.4.2 Industrial Ethernet

In this section, a general introduction into the technology of Industrial Ethernet as well as a delimitation to standard Ethernet is given. Consequently, migration concepts for incorporating the functionalities of traditional fieldbus systems into Industrial Ethernet applications are described, pointing out also the drawbacks and limitations of such purpose.

Between today's field devices and intelligent solutions for automation, a "connectivity gap" for seamless communication between the low-level automation devices and control solutions is still present. According to Pöschmann (2001), this "communication gap" describes the problem, namely that highly complex programming or parameterization is required in oder to access information from field devices throughout all levels of the automation pyramid (Jasperneite and Neumann, 2004). One major reason is a lacking continuous data exchange between different, heterogeneous communication systems that are present in various layers of an automation solution. At the same time however, field devices have been evolving from relatively primitive devices to cover a broader range of functions and subsequently become more naturally communicative (Jasperneite and Neumann, 2004; Ricot, 2001). Consequently, the tasks of automation devices are characterized by a shift from solely real-time behavior to more sophisticated tasks that involve historization, parameterization or diagnostic functions. These extended capabilities result in requirements that cannot be handled alone based on fieldbus protocols and communication systems. Thus, Ethernet-readiness is embedded in many new devices, however with the restriction of not loosing interoperability to existing fieldbus systems that are still required to guarantee real-time behavior in industrial processes.

One of the major aims that is connected to the introduction of Industrial Ethernet systems is to create an inherent intelligence being part of the computational network as every logical step in an industrial network has to be performed using intelligent behavior, either for processing physical signals to control actuators or to collect data from sensors (Gaj et al., 2013). Currently, human beings build up these sort of intelligently designed networking systems that are summarized under the term of Distributed Control System (DCS). From a conventional point-of-view, the aim of DCS is to continuously control industrial processes. This is traditionally being done on the basis of circuits that are dedicated to discrete control use-cases based on relays, and later PLC that represent devices somewhere between advanced relay circuits and "simple computers" (Gaj et al., 2013; Bolton, 2009). The state of art in creating intelligently behaving instances, however, is still far ahead.

Industrial Ethernet networking systems go one step further as they intend to enable an implementation of smart networking environment like they are already in use within conventional Internet or LAN topologies. Recent surveys concentrate on the sustainability of real-time capable Industrial Ethernet system (Danielis et al., 2014), especially in terms of:

Real-time capabilities Which performance and in that context what sort of real-time can be achieved in the best case?

Reliability Does the emerging network environment contain Single Point of Failure (SPoF)?

Scalability What is the maximum number of interacting devices that are able to exchange information through the real-time environment? This aspect is of special interest as the number of items to be connected within industrial contexts is going to increase substantially (Evans and Annunziata, 2012).

Self-configuration Is the system intended to perform self-configuration features such as dynamical adaptation to changes in the network topology overcoming the need for manual adaptations?

Hardware requirements Are future Industrial Ethernet systems able to operate on existing hardware configurations or is special hardware needed?

One of the most advanced and widely used Industrial Ethernet protocol is PROFINET, which stands for Process Field Ethernet (PROFINET) (Wellenreuther and Zastrow, 2011). PROFINET is characterized as an initiative to "emerge Ethernet to the next generation of industrial automation" (Feld, 2004). The PROFINET system consists of several functionalities such as modules for distributed automation (PROFINET CBA), decentralized field devices (PROFINET IO) as well as tools for the management of large networks, and integration approaches based on web application (Jasperneite and Feld, 2005). These additional modules intent to facilitate the adoption of standard Ethernet with its functionalities suitable for the automation domain (IEEE, 1998). According to Wellenreuther and Zastrow (2011) these PROFINET concepts include/enable the following functionalities:

- Industrial Ethernet networks with active networking components (switches, routers),

- Integration of existing fieldbus systems (PROFIBUS DP, INTERBUS, ...)

- Dedicated communication channels for automation specific transmission performance

- Cross-manufacturer engineering concepts (implementation, project development)

- Application and integration of IT infrastructures (Network administration, webservers, E-Mail, OLE for Process Control/Open Platform Communications/Openness Productivtiy And Collaboration (OPC), ...)

- Determined security in terms of EN 954-1/DIN EN ISO 13849-1 (DIN EN 61131-3, 2014) and IEC 61508 (DIN EN, 2010).

Despite the features of PROFINET or similar approaches, the introduction of Industrial Ethernet solutions alone does not mean that the industry is able to move easily from incompatible communication solutions to scenarios that are characterized by full interoperability of all Ethernet capable devices, as e.g. stated in Prytz (2008). In most cases, field level Ethernet based communication protocols such as PROFINET are not capable of communicating with the various industrial automation protocols that are located on top of Ethernet, i.e. the Industrial Ethernet protocol is not generically able to talk to an Ethernet/IP device. The solution to this issue are generally characterized by proxy devices, special intermediate components or by running numerous communication stacks simultaneously (Prytz, 2008), which does not represent a generic, scalable approach.

Even if these interoperability issues will get solved sufficiently, there is still shortages with regard to the responsiveness of such networks, especially in terms of applications that require hard real-time capabilities, such as devices for robotics, high performance PLC systems, motor control systems as well as instrumentation and I/O systems. All of these systems pose the same challenges on currently available Industrial Ethernet systems, which are rising requirements in terms of low latency, high update rates and high throughput (Prytz, 2008).

As a conclusion to the Industrial Ethernet systems it might be fair to say that the approaches introduced by the discussed concepts such as real-time extensibility of the standard Ethernet stack point to the right direction, especially due to the fact that modern Industrial Ethernet systems do not neglect the existence of emerged systems in automation. An incorporation of grown automation environment through proxies or similar technological approaches definitely is the way to go. However, despite the relatively generic interoperability to the lower levels of the automation pyramid, natural information exchange with higher information systems cannot be realized without further efforts.

The field of research in terms of Industrial Ethernet offers great potentials for tackling some of the main challenges of modern factories as pointed out in chapter 2. However, an exhaustive coverage of the abilities that are connected to the various Industrial Ethernet protocol families cannot be performed here, as this would go beyond the scope of this work. The interested reader might however refer to Appendix A.2, in which further basics and background information about the historical evolution of the Industrial Ethernet is covered.

3.4.3 Publish-Subscribe within Industrial Applications

The next step in pursuing natural interoperability between low-level systems and higher system architectures consists in the establishment of standards that are inspired by the idea of the IoT. The technologies that makes use of this ubiquitous computing paradigms are summarized under the term of Publish Subscripe (Pub-Sub) mechanisms.

The mechanism of publish-subscribe works according to a data-centric approach (Pardo-Castellote et al., 2005). One major characteristic of these approaches is that the exchanged information is in focus – not the underlying protocol. As a consequence, the information exchange is organized with regard to the data. This organization is realized in the form of *topics* that are used to group information that can be summarized according to a certain type of context, e.g. data from the same machine or data that has been collected at a specific point in time. The most common representatives of the communication pattern are *Message Queue Telemetry Transport (MQTT)*, *ZeroMQ* and *Advanced Message Queuing Protocol (AMQP)*. All of these standards are characterized by minimalistic protocol stacks in order to transfer messages as efficient and as fast as possible.

Another Pub-Sub protocol that is of special relevance for industrial applications is the Data Distribution Service (DDS). The key element of DDS consists in a combination of the publish-subscribe mechanism with additional features such as automatic discovery functions for creating ad-hoc network infrastructures and Quality of Service (QoS) definitions (Twinoaks Computing Inc., 2011). The QoS levels of DDS enable a specification of the data transmission mechanisms according to the requirements of the application, e.g. in order to transfer certain data sets with a specifically high performance in one case or with a focus on a high signal quality and reliability in another case (Pardo-Castellote et al., 2005).

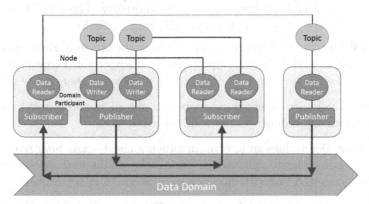

Figure 3.1: Data-centric information exchange using Pub-Sub (Pardo-Castellote et al., 2005)

Figure 3.1 shows the basic functionality of the data-centric design for realizing topic-based data transmission. The data domain can be characterized as some sort of "publish-subscribe cloud" the constitutes the backbone of the information exchange. In this context, the data domain is characterized a data-centric system bus mediating the data flow from all instances in the form of a message queue. The publishers and subscribers of the network are solely connected by means of a common "field of interest". A session or some other form of established connection between the publishers and subscribers is not necessary. The communicating parties do not even need to be aware of each other. The performance or other characteristics of the information exchange is determined via the QoS profile which each subscriber defines for himself. If the QoS characteristics of the publisher do not match those of the subscriber, no information exchange will be performed (Corsaro, 2016).

The MQTT protocol is another emerging Pub-Sub protocol that is also gaining more and more attention in industrial applications (Resnick, 2016). MQTT is a light-weight broker-based messaging protocol that was initially carried out for the application in SCADA systems (Obermeier, 2015). Originally developed by IBM and Arcom, MQTT is now an Advancement of Structured Information Standards (OASIS) specification that is standardized by means of version 3.1.1 (OASIS, 2014). The primary goal of MQTT is to facilitate the implementation of the IoT. One crucial requirement of MQTT is to operate within environments with limited resources, such as on limited computational power on embedded devices or in network environments with volatile or instable connection quality. The MQTT protocol is characterized by five basic features (Obermeier, 2015):

Last Will Clients are able to leave a message at the broker during connection initialization. If the client is improperly or disruptively disconnected, the broker will send the "last will" of the client to all subscribers in the network.

Sessions The broker saves the subscriptions of a client and is able to restore them if a client reconnects after a disconnection.

Keep Alive Determines an interval in which a client sends *ping* requests to the broker in order to validate the connection. This way, disconnections can be recognized faster than in comparison to TCP/IP timeouts.

QoS The Quality of Service levels for MQTT are graduated into three levels (see Figure 3.2).

The QoS level 0 offers the highest performance, yet less reliability. The data transfer is fast, but likely to fail. Duplicates data transfer is not possible though. QoS 1 guarantees that the message is at least transferred once, while QoS provides the highest reliability of data transfer by ensuring that each message is exactly delivered once.

The abilities of publish-subscribe mechanisms offer great potentials of becoming one of the next-level interoperability standards in manufacturing. The semantic scalability that is realizable by making use of communication channels might not be sufficient to map all requirements of low-level automation systems and their interoperability with high-level planning systems, but coupling with others semantic-rich standards could be a suitable complementation of establish interface standards like OPC Unified Architecture.

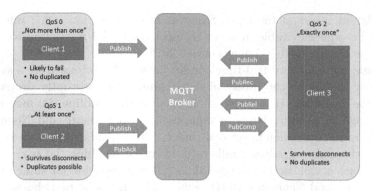

Figure 3.2: Quality of Service levels for the MQTT protocol

The full potential of the presented publish-subscribe mechanisms can only be achieved if the signal transport, which is usually realized by means of Ethernet, offers the bandwidth needed to guarantee real-time behavior. The next sections focuses on these advanced Ethernet solutions in terms of real-time Ethernet for time-sensitive networking.

3.4.4 TSN and Real-Time Ethernet in Industrial Applications

In the previous section, attempts have been pointed out that are suitable to bring near real-time behavior from shop floor to high-level systems of the enterprise. Recent advances in terms of Industrial Ethernet try to combine this high reactivity of Pub-Sub approaches with established Ethernet application by introducing an extended protocol stack.

Especially for companies that already make use of Industrial Ethernet solutions, a shift from combined fieldbus/Industrial Ethernet solutions towards a real-time Ethernet to connect devices from the shop floor is of high interest, technically as well economically. Thus, it is the aim of such companies to eventually replace conventional fieldbus systems completely by means of real-time capable Ethernet approaches (Danielis et al., 2014). Even though many attempts to adapt Ethernet for the usage within automation applications are initiated by industrial key players, there are still challenges to overcome. This is due to the fact that basic protocol technologies Ethernet relies on, namely Transmission Control Protocol (TCP), User Datagram Protocol (UDP) and IP, do not consider natural real-time capabilities in their communication stacks. Thus, in order to adjust standard Ethernet for the usage with real-time applications, several adaptations and modifications on different layers of the Open Systems Interconnection (OSI) reference model become inevitable in order to meet industrial restriction for hard real-time requirements (Danielis et al., 2014).

In this context, hard real-time conditions are characterized by a system behavior in which "the utility of data transmission abruptly drops to 0 when the deadline (D) is exceeded. Exceeding the deadline (D) may result in damage to product parts or equipment" (Jasperneite and Neumann, 2004). On the other hand, soft real-time conditions are defined as a state in which "the utility gradually falls to 0 when the deadline (D) is exceeded. In this case a violation of the deadline (D) is acceptable for some events" (Jasperneite and Neumann, 2004). Effectively, this definition characterizes hard real-time constraints as the strongest form of time constraint that can be present in terms of automation applications. According to this definition, data becomes useless immediately after failing a certain deadline.

To meet the hard real-time conditions that are required in some automation applications, e.g. motion control, in which the deadline is usually below 1 ms, a multitude of real-time capable Industrial Ethernet systems had been developed. This developments intend to solve problems of standard Ethernet- and TCP/IP or UDP/IP-based communication by means of various approaches. These approaches have in common that they aim at improving the abilities of automation systems with regard to real-time capabilities, reliability, scalability, self-configuration of the network, and restricted hardware requirements (Danielis et al., 2014).

The first sophisticated attempt to establish real-time behavior within industrial application has been initiated by the Institute of Electrical and Electronics Engineers (IEEE) Task Group for TSN in terms of IEEE 1722-2011 (2011); IEEE 802.1BA (2011); IEEE 802.1Qat (2010); IEEE 802.1Qav (2009) and IEEE 802.1Qbv (2015). The introduction of TSN hereby attempts to facilitate the migration from industrial communication based on proprietary bus systems to a further expansion of Industrial Ethernet solutions. The basic model stack of TSN has already been applied by various stakeholders, e.g. by means of *Ethernet/IP*, *Modbus TCP* or *Foundation HSE*. These application do not guarantee hard real-time behavior so far, however, according to Fröhlich et al. (2013), the real-time capabilities of these applications are sufficient in most use-cases. According to the IEEE, however, the final version of the TSN standard intends to guarantee hard real-time capabilities under all circumstances through reservation and prioritization mechanisms (IEEE 802.1Qav, 2009).

The last point is of major importance, especially in applications that rely on deterministic real-time behavior, high reactivity and strong synchronization in recurring processes. For such applications, time frames of less than $1\,ms$ are often required, e.g. in motion control or robot interaction. Earlier attempts to reach real-time behavior for these sort of applications relied on proprietary solutions and were carried out in the form of extensions for *Profinet IRT, EtherCAT* or *Sercos III*. The IEEE TSN standards concentrates on low-level improvements of the Ethernet stack as on proprietary solutions to reach real-time behavior. The IEEE focuses on time synchronization for network nodes/hosts and on the development of software building blocks to enable resource reservation for critical data streams.

The standardization around TSN offer great potentials to play a major role in future automation systems and to ultimately replace proprietary fieldbus solutions on the shop floor. The subject is also of special interest in this work as recent developments of the TSN and Pub-Sub communities focused on the incorporation of real-time Ethernet and publish-subscribe mechanisms into sophisticated interface and communication standards such as OPC UA. The coupling of these technologies stacks will be of special interest in later section of this work, the incorporation of OPC UA with Pub-Sub/TSN will be covered in section 3.9.7.

3.5 Web Services and Service-Oriented Architectures

Apart from the transport systems that are needed to transfer data among automated production systems, means to realize this transfer in distributed, loosely-coupled systems are of vital importance. Especially, in terms of establishing CPS in manufacturing environments, the usage of web services is inevitable for the realization of a scalable information exchange, e.g. in order to increase the production flexibility (Jammes and Smit, 2005).

Thus, in addition to a language that describes information and provides data syntax and semantics, a common underlying mechanism for carrying the information from one entity to another or to perform interactions is needed. In this context, a web-based infrastructure offers potentials to significantly increase the interoperability and ubiquity in these systems (Lastra, 2006). The semantics of these services should comply with certain model definition, e.g. based on the Extensible Markup Language (XML) and Web Ontology Language (OWL) mark-up standards in order to enhance the services with explicit semantics.

The introduction of Web Service (WS) and SOA immediately address the aforementioned requirements (Huhns and Singh, 2005). In this context, WS are defined as "self-contained, self-describing, modular application that can be published, located, and invoked across te Web" (Tidwell, 2000). The generality of this definition implies that WS "provide an interface for encapsulating a process, any process, whether a business process, a software process, an interaction process, or a physical process. Factory or production processes are intrinsically included in this definition." (Lastra, 2006)

One important feature that makes WS especially relevant for an application in automation scenarios is its capability to compose complex processes in a distinctive order. As a typical feature of manufacturing systems consists in processes that are composed and executed in certain sequences and process orders, a composition of processes is of special importance when establishing WS in a factory environment. By means of WS it is possible to encapsulate these complex orders of processes into a single service definition (Lastra, 2006). The practical creation of sequences and based on a synchronized execution of services that encapsulate a number of distinctive process is referred to as *orchestration* (Jammes et al., 2005). The orchestration hereby implements the application logic that is needed in order to carry out atomic processes and at the same time provide a high-level web service interface for the composed process Lastra (2006).

Another concept that can be regarded as complementary supplement to the establishment of flexible web services is the concept of service *choreography* (Peltz, 2003). While the process of orchestration solely focuses on the execution of atomic services in a certain sequence, orchestration does not take into account different types of conversation patterns that are required in order to evoke the atomic processes. Choreographies on the other hand precisely take into account messages and interaction sequences that are required to execute a service through a certain web service interface (Lastra, 2006). Thus, choreographies do not only focus on the atomic processes, but also on means to realize the triggering of the encapsulating web services in a SOA. This sort of harmonization of services and atomic processes is of vital importance in terms of distributed environments. In the context of this work, similar approaches need to be taken into account with regard to specialized SOA approaches in terms of multi-agent systems and their communication through generic interface standards.

Although the application of WS and SOA is not directly targeted in this work, it should be mentioned that the approaches which are used to implement the proposed framework, namely multi-agent systems and semantic communication trough OPC Unified Architecture, both make use of web services or are at least applicable of an incorporation with WS architectures, e.g. in terms of an Enterprise Service Bus (ESB) or similar realizations of SOA.

3.6 Intelligent Production Automation by Means of Multi-Agent Systems

This section focuses on agent-based approaches applied to modern production. Regarding a general understanding, the term agent by means of an autonomous entity within a manufacturing environment is defined according to Jennings and Wooldridge (1998) as follows: "an agent is a computer system situated in some environment, and that is capable of autonomous action in this environment in order to meet its design objectives." Accordingly, agents should be able to interact with other agents or in contact with human beings, and thereby should be in charge of their own actions and internal state (Shen and Norrie, 1999).

3.6.1 Fundamental Concepts of Multi-Agent Systems

An MAS consists of multiple intelligent and interacting entities called agents, which are embedded within a determined environment. The goal of MAS is to make use of its individual intelligent agents in order to fulfill a task that an individual, traditional system could not perform or not fulfill as efficiently (Wooldridge, 2009).

As pointed out by Jennings (2000) there is a distinctive difference between MAS and other distributed computer systems, although both concepts show certain similarities. To clarify these differences, the concept behind an agent need to be specified in a proper way. According to the basic definition of the agent term mentioned above the main difference between agent-based approaches and similar systems of distributed computing environments consists in the autonomy of an agent in contrast to the service paradigm addressed by distributed environments, in which each entity represents a callable service omitting any autonomous reaction from his own part.

In order to achieve autonomy, an agent needs to be in possession of its own design objectives aiming at the goals and target states of these objectives with every actions. In addition to autonomous behavior, an agent entity also needs to be capable of three additional requirements to be qualified as an agent in terms of MAS (Wooldridge and Jennings, 1998):

Reactivity Agents need to be aware of their environment and must be able to react to external changes within a proper time range, if a reaction on the part of the agent is necessary to fulfill its design objectives. Thus, both the environment and the agent are able to influence each other in terms of system state changes in the environment and autonomous actions of the agent. In contradiction to the function of distributed computing system, the environment cannot trigger single services through an agent, thus can not *force* the agent to perform a certain actions. The decision whether the agent will perform a (necessary) action always remains with the agent. Furthermore, agents are also capable of deciding, in which way they will react on certain changes in its environment.

Proactiveness An agent is always encouraged to act goal-oriented with regard to its design objectives. Solely reacting to changes within the environment may not be sufficient to reach the goals of the agents' objectives. Therefore, besides autonomous and reactive behavior, agents also need to act in a proactive way. In order to enable this proactiveness behavior, an agent has to be in charge of knowledge about the impact of its actions. This knowledge might exist in the form of predefined rules or in terms of algorithms that enable learning and in this way evolve its knowledge on the basis of expert systems and artificial intelligence.

Social ability The social ability of an agent describes its capability to interact with other agents. The interaction capabilities between agents are comprised of information exchange as well as triggering of actions. The information that is exchanged between agents attempts to complement the world view of the agent in a complete way. The demand for actions from one agent to another is – as always – not binding, thus an agent can never be forced by another entity to perform a certain task and accordingly decides autonomously about the task and the task's suitability for its design objectives.

Especially the proactive behavior of agents will be of major interest in later parts of this work, as the evolving of agents in current MAS is often limited by certain boundary conditions of the agent system, e.g. in terms of scalability or communication capabilities. The approaches that are targeted in this work aim at enabling a fully proactive behavior of the agents based on scalable learning functions and indefinite communication potentials.

In practical applications of MAS, each agent is designed with regard to a special purpose. The purpose of an agent is of major importance for the definition of its design objectives, which comprise the goals and general targets of the agents' actions. However, in addition to the design objectives, the purpose of an agent is also decisive for the determination of its capabilities, which describe the set of actions an agent is capable of. A typical MAS consists of multiple agents with different purposes, design objectives and abilities. Accordingly, the fulfillment of a complex task that involves many of the available capabilities, requires collaboration of the agents involved. For this sort of collaboration all agents need to be capable of communicating with each other in an integrative way, i.e. the information processing concepts of an agent has to match those of the other agents in order to understand information in a way that is common to the other communication partners. According to Wooldridge (2009), this type of communication between agents can be realized if: (i) all agents are able to find and identify each other and (ii) all agents make use of a message system with a predefined ontology, which every agent is able to understand.

Especially the first point of prerequisites requires a proper description of each agent. Specifically, in terms of enabling an expedient collaboration between the agents, the following conditions have to be met by the description features of an agent:

- Agents need to be able to search for other agents within the MAS, i.e. agents have to provide a specific endpoint, for example in a network environment, to be identifiable as an agent.

- Furthermore, agents need to be able to read the description of each other in order to get a proper picture of their surrounding community.

- The description of the agent has to contain the basic design objectives of an agent or at least its capabilities, because agents need to be aware of the basic goals and/or abilities of their community

Agent Platform

To provide a proper description of an agent that is readable, understandable and interpretable by other agents in an integrative manner, the description model of each agent needs to follow common design principles, e.g. by making use of a common ontology description, fixed namespaces for agent capabilities, etc. One of the most important attempts to meet this challenge of standardizing the descriptive model of an agent has been carried out by the Foundation for Intelligent Physical Agents (FIPA), which is a non-profit organization founded in 1996 with the explicit purpose to define design principles to standardize the way agents are described and how they communicate. The basic structure that has been carried out in the course of these endeavors is represented by the Agent Management Reference Model, also known as the FIPA Agent Platform Reference Model (Foundation for Intelligent Physical Agents, 2002b). The model attempts to define a basic context for enabling agent collaboration and communication through the definition of additional components for MAS and their respective responsibilities. The structure of the model is depicted in Figure 3.3.

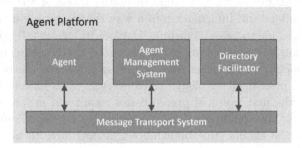

Figure 3.3: The Agent Management Reference Model by FIPA

This abstract representation of an MAS shows the basic components current state-of-the-art agent-based systems are composed of. The *Message Transport System* represents the facilities that enable interconnection or communication between the different components of the agent platform. The *Agent* represents an arbitrary number of smart agents being part of the MAS, the *Agent Management System* summarizes the administrative capabilities of the agent platform while the *Directory Facilitator* keeps track of the capabilities that are present through the available agent.

Agent Management System

Although multi-agent systems in their original intention are characterized as decentralized systems made up of autonomous agent entities, the central component represented by the *Agent Management System (AMS)* is needed to ensure the availability of certain services such as *discovery functions*, i.e. the agents need to be able to find each other by means of some backbone structure/architecture, or similar administrative matters. The environmental structure of an MAS is usually managed by means of exactly one AMS ensuring the management of centralized services. All available agents that intend to become active members of the agent platform, i.e. the MAS, have to register to the AMS in order to be recognizable as available entities inside the multi-agent infrastructure. In the common literature, the services that allow for finding and identification of other agents are referred to as *white page services* (Foundation for Intelligent Physical Agents, 2002b).

Directory Facilitator

In order to enable a suitable collaboration between the agents of an MAS every agent needs to know about the abilities of other agents within their environment. In conventional MAS architectures this is realized in two steps: (i) Firstly, the agent that is interested in its surrounding entities obtains a list of available agents through the AMS; (ii) in a second step, the agents asks every agent about its capabilities. Although this procedure fulfills the agents' target, it is very time consuming as it brings along a tremendous overhead in terms of information that need to be exchanged. Especially in MAS consisting of a large and fluctuating number of agents, this procedure is not efficient as each agent need to refresh its entire list of known abilities prior to each production step or negotiation action in order to be aware of its environment (Wooldridge, 2009).

Thus, the Directory Facilitator (DF) is another component that is of critical importance for MAS as it offers services that automatically keep track of the abilities offered by each agent. The smart agents that are available within the agent platform can make use of these services in order to find out about available capabilities within their environment. Using this knowledge agents are accordingly able to split work packages and to negotiate collaboration scenarios within their community, while at the same time reducing unnecessary information exchange with other agents. Thus, although the DF is an optional component of the agent platform, its presence is inevitable for the management of an agent system that intends to react to environmental changes in reasonable time ranges. Accordingly, agents should register their abilities with the DF in such way that the DF is always able to provide an up-to-date list of available capabilities. The services that are related to information exchange between agents and the DF are summarized under the term of *yellow page services*. By making use of this service the load of messages exchanged between components in the MAS can be reduced significantly as it requires just one message to and from the DF to check available capabilities instead of messages between all agents. (Foundation for Intelligent Physical Agents, 2002b)

Messages between Agents

The communication between agents that has been mentioned above as some sort of abstract mechanism to exchange information between different entities of the MAS is further specified in this section. Communication between agents is needed in terms of:

- Exchange of internal agent or external environmental data or information,

- cooperation in terms of shared plans, abilities and targets,

- negotiation techniques to cooperatively solve problems or to compete for available tasks.

All communication between the agents is hereby realized by means of messages. The utilization of messages as the prior communication channel is favored in comparison to other exchange methods such as the usage of callable services as the usage of messages preserve the autonomy of each agent. Thus, if an agent receives a message, the agent can either decide to reply to the message, to perform a demanded task or to ignore it. Within a determined agent environment, such as an MAS, messages are directly sent from a sending agent to the receiver agent. In order to send messages to locations outside the MAS the FIPA specifies the *Message Transport System* for the message exchange e.g. between different MAS.

In order to be interpretable to other agents the messages need to comply to a certain set of common semantics. In terms of the standardization regarding messages exchanged between agents, the FIPA specifies an ontology that is called Agent Communication Language (ACL) (Foundation for Intelligent Physical Agents, 2002a). The basic structure and representation of such messages is depicted in Listing 3.1. The source code shown in Listing 3.1 represents a message from client named *dummy* that requests a registration at the AMS. The structure of the ACL message format intends to be readable for machines as well as easily understandable by humans. Thus, the ontology that defines this basic structure makes use of common English words and is therefore intuitively interpretable. As the basic structure of the ACL message is independent of the message's content this simple example message contains all necessary elements for the generation of arbitrary ACL messages. The structure is accordingly made up by the following main elements:

Listing 3.1: Standardized ACL message format

```
 1  (request
 2      :sender
 3          (agent-identifier
 4              :name dummy@foo.com
 5              :addresses (sequence iiop://foo.com/acc))
 6      :receiver (set
 7      (agent-identifier
 8          :name ams@foo.com
 9          :addresses (sequence iiop://foo.com/acc)))
10      :language fipa-sl0
11      :protocol fipa-request
12      :ontology fipa-agent-management
13      :content
14          "((action
15          (agent-identifier
16              :name ams@foo.com
17              :addresses (sequence iiop://foo.com/acc)
                )
18          (register
19              (ams-agent-description
20                  :name
21                  (agent-identifier
22                      :name dummy@foo.com
23                      :addresses (sequence iiop://foo
                        .com/acc))
24                  :state active))))")
```

Performative The performative defines the type of the present message and at the same time represents its purpose, i.e. the internal meaning of the message. In terms of the message shown in Listing 3.1 the purpose of the message is a *request.*

Receiver and sender The receiver and sender elements of the message define the participating, i.e. communicating or collaborating entities that are involved in the information exchange. A precise allocation of each agent is realizable through its globally unique identifier, e.g. a name. Within the example message, an optional parameter, the address of the agent is added to the sender and receiver elements.

Language and ontology These parameters shown in the language and ontology section of the message further describe the content structure of the message. In terms of the ACL specification different formal languages and ontologies can be applied to characterize the message content and to offer a formal description for the semantic parsing/understanding of the according message.

Protocol The protocol defines the interaction/exchange protocol that is used for the transfer of the message.

Content The content finally represents the actual payload of the message. In Listing 3.1 the agent identified by the name of "ummy@foo.com" performs an action that intends to register at the abstract agent "ams@foo.com" that represents the AMS and changes its *state* to *active*.

As seen in the formal description of the message body the structure of the message is standardized in a distinctively strict manner. The assembly solely allows for information exchange between two agents. Accordingly, even the AMS as the management and coordination center of the MAS is interpreted as an agent in order to allow for a message exchange between the agent and the AMS.

This concept already offers some weaknesses regarding this sort of message transfer. Although the definitions determined by the FIPA constitute a robust way to exchange simple messages, the semantics and scalability characteristics of the according messages seem to be rather weak. Thus, in order to reflect complex interaction plans between the agents, messages become complicated or will even not be capable of representing the desired content. Especially this point is a target objective of this work and will be further addressed in the introduction of the semantically scalable interoperability framework derived in chapter 5.

3.6.2 Communication Protocols and Standards for Multi-Agent Systems

The transfer of messages such as defined by the FIPA ACL standard requires the utilization of an exchange protocol that takes care of the physical transportation of the payload and meta information. As agent-based approaches are carried out according to the paradigms of SOA the goal of most agent-oriented solutions is to provide flexible means for realizing a practical message exchange. In the current literature, web services as well as solutions inspired by SOA are characterized as key issues for the realization of agent-based solutions (Leitão et al., 2016), see for example Leitão and Karnouskos

(2015), Karnouskos and Tariq (2008), Karnouskos and Tariq (2009), Ribeiro et al. (2008) and Mendes et al. (2009). In terms of these SOA-based approaches, especially the Device Profile for Web Services (DPWS) standard as well as the concept of Representational State Transfer (REST) are worth to mention. DPWS which was defined by the OASIS deals with matters of ubiquitous devices integration using web service approaches embedded into distributed devices. In this context, REST can be regarded as an alternative integration approach focusing mainly on Machine-to-Machine (M2M) and emerged out of the world wide web initiative. (Leitão et al., 2016)

The FIPA already provides certain transport protocols and proposes the Hypertext Transfer Protocol (HTTP), Wireless Application Protocol (WAP) and Internet Inter-Orb Protocol (IIOP) protocols for the exchange of messages between agents (Bellifemine et al., 2007). As these protocols are widely used for the application in web based systems such as the Internet they can be referred to as web protocols. The usage of web protocols within this context can provide several advantages, e.g. due to the (i) well-defined format of messages and the (ii) exceedingly high prevalence of these web protocols in all kinds of applications.

With regard to the realization of real-time applications and control tasks that are inevitable within industrial environments, some further standards had been made available for the usage with MAS (Leitão et al., 2016):

- The widely distributed and well known standards for the programming of PLC – IEC 61131 – and its according modules for the development of real-time solutions for process automation through industrial computers and controller devices. An integration of IEC 61131 standards furthermore requires the implementation of execution models for cyclic and event-driven execution of control tasks (Otto and Hellmann, 2009). Furthermore, several domain-specific programming languages had been made available through the IEC 61131 standard.

- An additional standard that is of major interest within the domain of industrial automation is the IEC 61499 (2000) standard defining a function block oriented reference model for distributed, reconfigurable automation and control systems (Leitão et al., 2016; Vyatkin, 2011). In comparison to the IEC 61131 standard the function block oriented IEC 61499 can be characterized as a pure event-driven execution approach that pursues especially targets of distributed control requirements in terms of (re-)configurability, portability and interoperability in a distributed control architecture.

- Another integration solution for field level devices was introduced with the IEC 61804 standard (IEC 61804, 2015). The specification provides means for enabling a generic interface to field level devices by following the Electronic Device Description Language (EDDL), an extended version of the Device Description Language (DDL) that was carried out in cooperation of three major players in the field of shop floor automation – the HART Communication Foundation, PROFIBUS and the Fieldbus Foundation.

Especially the standards that focus on the integration of low-level devices and automatic control solutions such as IEC 61499 will be of special interest in this work and will be further discussed in the derivation of a reference architecture based on MAS in chapter 4, section 4.2.2.

The realization of the communication between distributed devices in terms of CPS or agents is carried out by the usage of established approaches (Leitão and Karnouskos, 2015; Colombo et al., 2015) such as the Extensible Messaging and Presence Protocol (XMPP) that proposes an XML-based format for the realization of message-oriented middlewares. The adaptation of these standards is especially advanced in the domain of smart grids (intelligent electrical networks) and used as a standard transport protocol as part of the IEC 61850 standardization (IEC 61850, 2013).

Major drawbacks in terms of using the standards mentioned above, however, also have to be pointed out, especially in terms the requirements of Industry 4.0 applications. The downside of using solely web-based protocols for the communication between agents are:

- The lack of means to support encryption of messages. Although the communication channels for the application of web-based protocols might be secured, the messages, i.e. their payload might be readable and interpretable by unauthorized third parties.

- Authentication mechanisms in terms of access rights or other permission related privileges are not properly defined for these protocols.

Possible other communication exchange could be realized by means of the discussed Pub-Sub protocols that intend to enable real-time communication between peers based on a broker or communication channel model. These solutions have been applied to several MAS application. Worth to mention in this context are the MQTT and the Simple Object Access Protocol (SOAP) technologies enabling light-weight M2M communication by making use of TCP. However, most of the publish-subscribe and IoT inspired approaches in use do not reflect the system complexity as it is not possible to map the richness of available meta data and contextual information using channel and/or certain publishing addresses.

Finally, the OPC UA standard presented in IEC 62541 (2015) is a major harmonization approach to harmonize the information exchange in industrial systems (Leitão et al., 2016; Leitner and Mahnke, 2006). Applications of OPC UA and DPWS have been already demonstrated in the connection of CPS within industrial environments (Colombo et al., 2015; Karnouskos et al., 2014).

With regard to the available protocols and communication solutions currently deployed for the realization of CPS/MAS the most holistic and scalable approach seems to be present in terms of the OPC UA standard. Thus, especially this standard will be the major choice for further development with regard to the desired reference architecture for MAS with high interoperability and scalability from the shop floor to the high-level information systems. A multi-agent system approach based on OPC UA as the backbone for communication and for information modeling will be presented in chapter 5.

3.6.3 Agent-based Approaches in Production

After pointing out the technical background of MAS as well as common standardization in terms of their realization, this section will focus on existing solutions that have been applied to current production systems.

As already pointed out in the motivation and description of fundamentals, global competition and rapidly changing requirements driven by various customer demands are forcing changes regarding production styles and the general configuration of manufacturing organizations (Shen and Norrie, 1999). Moreover, traditional methods that focus on centralized solutions for planning, scheduling and control mechanisms are characterized as insufficiently flexible in order to respond to dynamic variations in production requirements. These traditional approaches furthermore lead to high risks in terms of single point of failures and similar problems that are likely to shut down entire systems. The agent paradigm is one possible idea to overcome these issues in a natural manner by introducing autonomous independent subsystems that implement a distributed intelligent manufacturing environment.

Especially agent-based solutions are capable to face the issues mentioned above. The ideal picture draws future factories that are able to deal with these issues and will be flexible in terms of global competitiveness, new product innovation and introduction as well as rapid market responsiveness. The underlying next generation manufacturing systems that provide means to realized visions is able to work strongly time-oriented, and at the same time focusing on cost and quality. The requirements of these manufacturing systems are met by the idea of agent-based approaches and can be summarized as follows (Shen and Norrie, 1999):

Enterprise integration Summarizing the challenges and tasks connected to an integration of management systems such as purchasing, orders, design, production, planning & scheduling, control, transport, resources, personnel, materials, quality as well as the communication with partners of the external supply chain.

Distributed organization Characterizes the demand for knowledge-based systems across distributed organizations that are needed to link the management of customer demands directly to resource and capacity planning and scheduling.

Heterogeneous environments Demands for the incorporation of heterogeneous hardware and software solutions that bridge applications between the manufacturing and information management parts of the enterprise.

Interoperability Further characterizes the heterogeneity within information environments in terms of different programming languages, data representation and models that enable the operation in different computing platforms. Interoperability requirements aim at an integration of these different systems by enable translation engines or suitable interfaces.

Open and dynamic structure It is inevitable to enable running system for a generic extension during operation. This, it must be possible add subsystems dynamically without the need to stop or reinitialize the system.

Cooperation Manufacturing enterprises need to fully cooperate with partners of the supply chain, such as suppliers, partners and customers for commercialization. Such cooperation need to be carried out efficient and in a quick-response manner.

Integration of humans with software and hardware Human begins need to be integrated with hard- and software. Information from heterogeneous sources should be accessible and information exchange need to be realized in a bi-directional way.

Agility The adaptation of product life cycle need to be reactive to customer demands and must support a quick reconfiguration regarding continuous and unanticipated changes.

Scalability The incorporation of additional resources into the organization for an increase of the production should be realizable without disrupting the organizational flow.

Fault Tolerance The manufacturing system needs to be resilient to faults in the production and should compensate these faults by means of a quick recovering.

Ahead of the various initiatives focusing on the usage of CPS in the field of industrial manufacturing, several important research projects have shaped the integration of agent-based approaches with CPS ideas. Regardless on the increasing popularity of the CPS term especially in the previous couple of years, these former research projects have paved the way for the intensive research programs mentioned above. The most important projects including their impact on the different levels of the ANSI/ISA 95 standard are shown in Table 3.1 (Leitão et al., 2016), which gives a general overview regarding general applications of agent-based approaches in *Smart Production.*

3.6.4 Agent-Based Solutions for Intelligent Manufacturing

Artificial intelligence by means of agent-based solutions have been applied to manufacturing for quite some time. The first attempts to integrate agent approaches to modern manufacturing go back to the beginning of the 1990s and focus on applications that take into account the requirements for next generation manufacturing systems mentioned above such as manufacturing enterprise integration, supply chain management, manufacturing planning, scheduling and control, materials handling as well as holonic manufacturing systems. The methods that are being applied along with these agent technologies lead to the a number of key issues in developing agent-based manufacturing systems, which Shen and Norrie (1999) characterizes as *Agent technology for enterprise integration and supply chain management, agent encapsulation, multi-agent organization and system architectures, dynamic system reconfiguration, learning in agent-based manufacturing systems, design and manufacturability assessments, distributed dynamic scheduling, planning & scheduling & execution, factory control architectures* and finally *tools and standards.*

Table 3.1: Applications of agent-based approaches in Smart Manufacturing according to Leitão et al. (2016)

Name	ISA 95 Level	Scope	Year	Source
BHP Biliton	L2	Process control	1995	Mařík et al. (2005)
Yokogawa	L1	Machinery control	1998	Wada et al. (2000)
MASCADA (Mercedes Benz)	L2	Manufacturing control	1999	Brückner et al. (1998)
Cambridge packing cell	L2	Manufacturing control	2003	Mönch et al. (2003)
FABMAS	L2	Process control	2003	Mönch et al. (2003)
PABADIS	L1–L4	Manufacturing control	2004	Lüder et al. (2004)
ABAS	L2	Manufacturing simulation and control	2005	Lastra et al. (2005)
SO-CRADES	L1–L4	SOA-ready devices, cross-layer integration (ERP)	2007	Karnouskos and Tariq (2008, 2009); Colombo et al. (2015)
ADACOR-FMS	L2	Manufacturing control and reconfiguration	2008	Leitão and Restivo (2008)
IDEAS	L2	Reconfiguration and plug & produce	2013	Ribeiro et al. (2015)
ARUM (Airbus)	L3	Production planning and scheduling	2015	Marin et al. (2013)
ADACOR2	L2	Manufacturing control and reconfiguration	2015	Barbosa et al. (2015)
PRIME	L2	Manufacturing plug & produce	2015	Rocha et al. (2014)

These issues are in focus of many research activities and projects that have been carried out since the early 1990s. Future agent-based applications are strongly encouraged to define concepts that deal with these issues in some way. Issues that are interest in this work are especially *multi-agent organization and system architectures, dynamic system reconfiguration, learning in agent-based manufacturing systems* as well as *planning & scheduling & execution*. A distinctive literature review about these topics is carried out in Shen and Norrie (1999) and Leitão et al. (2016).

Research activities involving the application of agent-based solutions to smart manufacturing in terms of planning, scheduling and control are of major importance regarding agent-oriented research subjects. The organizational steps of planning, scheduling and control are also a vital objective of this work as these subjects address all of the mentioned hierarchical levels of the automation pyramid. In order to focus on these low-level activities of agent-based approaches the next section will concentrate on Holonic Manufacturing Systems that characterize smart entities on the shop floor more from the manufacturing resource point-of-view.

3.6.5 Holonic Manufacturing Systems

In contrast to agent-oriented technologies like MAS, the development of Holonic Manufacturing Systems (HMSs) is more focused on the manufacturing domain, addressing specifically the needs of an evolving manufacturing system. The research that has been performed in terms of HMS accordingly concentrates on the challenges of modern production.

The concept of HMS differs from the definition of solely computer science related concepts as the term holon is of philosophical nature. First mentioned by Arthur Koestler (Koestler, 1989), the word *holon* is a combination of the Greek terms *holos* which means "whole" and the suffix *-on* expressing that the holon is a particle or part of another system. Figuratively speaking this means that a holon as a whole is part of something bigger, i.e. a system architecture or similar structure. In terms of manufacturing systems, two central concepts regarding holons can be pointed (van Brussel et al., 1998):

- Complex systems evolve from simple subsystems if there are stable intermediate forms. The resulting system configurations will be of hierarchical nature.

- The holon term points out the hybrid nature of *sub-wholes*/parts in real systems. In this sense holons are self-contained wholes to their

subordinated parts, and dependent on other parts when regarding from
the other direction.

Regarding manufacturing systems, the target is to make use of "the benefits
that holonic organization provide to living organizations and societies, i.e.
stability in the face of disturbances, adaptability and flexibility in the face
of change, and efficient use of available resources" (van Brussel et al., 1998).

The term holonic manufacturing was first mentioned in the beginning of
the 1990s and has drawn a lot of academic attention in industrial research
(Bussmann, 1998). The development of HMS was initiated along with
research on a number of other disruptive manufacturing paradigms (van
Brussel et al., 1998) such as bionic manufacturing (Okino, 1993), genetic
manufacturing (Ueda, 1993), the fractal factory (Warnecke, 1992), random
manufacturing (Iwata et al., 1994) and virtual manufacturing (Kimura,
1993). The paradigm for holonic manufacturing was carried out as part of
the framework of the Intelligent Manufacturing Systems (IMS) programme.
The HMS was one of six test cases that had been carried out in terms
of a project aiming at research about system components of autonomous
modules and their distributed control (van Brussel et al., 1998). Similar to
the goals of agent-oriented approaches, HMS aim at better understanding
the requirements of future-oriented manufacturing systems and to figure out
ways to reach the prerequisites of such production systems. The determined
goals of a holonic manufacturing architecture is to "enable (easy) self-
configuration, easy extension and modification of the system, and allow
more flexibility and a larger decision space for higher control levels" (van
Brussel et al., 1998).

HMS are developed closely to the manufacturing process and have been
applied to multiple research areas in the field such as flexible shop floor
control (Valckenaers, 1993), non-linear process planning (Detand, 1993)
and reactive scheduling (Bongaerts et al., 1995). Based on these complex
planning methodologies along with machine controllers, multiple testbeds
were carried out as prototypes for functional HMS (van Brussel et al., 1998).
One of these testbeds consists of a flexible assembly system and has been
implemented as a holonic shop floor control architecture (van Brussel et al.,
1994). According to van Brussel et al. (1998) the HMS concept combines
the best features of hierarchical and heterarchical organization (Dilts et al.,
1991). A heterarchy is an organizational structure, in which the elements of
the organization are unranked (non-hierarchical) or where they possess the
potential to be ranked a number of different ways (Crumley, 1995). There

are three main requirements of HMS (Holonic vision of autonomous and cooperating entities Botti and Giret (2008)):

- "Firstly, holons are entities with autonomous control over the machine behavior they are associated with. Holons may create and execute their own plans and follow their own strategies. This autonomous behavior implies some kind of decision-making component that guides the holon physical control." (Botti and Giret, 2008)

- "Secondly, two or more holons are able to cooperate when and wherever it is necessary. To do this, these holons are able to figure out cooperation opportunities, make cooperation or negotiation commitments, and finally execute the cooperation committed to." (Botti and Giret, 2008)

- "Thirdly, holons are able to act in multiple organizations called holarchies and these holarchies are created and modified dynamically. Creating a holarchy means to aggregate the manufacturing process or the controlling process in order to enhance productivity. This implies work and responsibility distribution, and the definition of interaction patterns, which means that holons are able to figure out opportunities for reorganization, negotiate reorganization, and follow the interaction pattern." (Botti and Giret, 2008)

The fundamental insights about HMS in this section will be of major interest in chapter 4 as the proposed architecture contains elements from both MAS and HMS approaches.

3.6.6 Comparison of Traditional and HMS Inspired Control Solutions

In the light of HMS, the control of manufacturing systems can be handled in a different manner. Specifically, the conceptual delimitation of HMS in contrast to traditional manufacturing control solutions is motivated by some fundamental drawbacks (Bussmann, 1998):

- The inability of existing manufacturing systems to deal with continuous evolution of products within existing production facilities and

- insufficient means to maintain satisfying performance outside the range of standard operation conditions.

Especially with regard to the well established CIM technologies, the main differences of a flexible manufacturing organization in terms of agent-based/HMS approaches can be pointed out. In contrast to these HMS inspired approaches, CIM technologies have been characterized as inflexible, fragile and difficult to maintain (McFarlane, 1995):

- CIM based systems are characterized by a fixed control hierarchy and do not support changes in the system configuration and interdependencies between the different layers.

- Reconfigurations and extension of present systems can not be performed easily.

- Production performance can not be maintained outside of normal operation conditions.

- The access of low-level information like machine data is difficult to perform.

- Control operations are completely automated and lack of human intervention points.

In contrast to these rigid communication hierarchies and top-down organization flows, the HMS based approaches aim at organizing the manufacturing process in a more flexible manner. According to McFarlane and Bussmann (2003), holonic control solutions are specially qualified to function independently of any particular products, to dynamically change hierarchies of control functions and to involve distributed problem solving in dynamic control strategies.

Thus a combination of agent-based an holonic manufacturing system will serve as the best starting point for the development of framework that enables learning agent in a distributed environment. The evolving of these agents in such combined approach will be in focus in chapter 4, section 4.4 and chapter 7.

3.7 Machine Learning in Distributed Environments

Although the main focus of this work is on the information modeling of decentralized automation systems, an extension of single agents in terms of machine learning will be carried out. Additionally, an evaluation use-case for an emerging multi-agent system will be given in terms of a machine learning scenario. This section contains the fundamentals concepts of the machine learning techniques that will be applied in chapter 7 and in chapter 8, section 8.2.

3.7.1 Basic Characteristics of Machine Learning

The terms Machine Learning (ML) and Artificial Intelligence (AI) are often used synonymously. As a matter of fact, both fields of research share common ground as they are both located in the research area of computer science and are mostly used in combination. However, the area of ML should rather be characterized as a subset of AI just like the paradigms of intelligent agents are also summarized under AI methods. A further distinction of ML and AI can be carried out according to the following definitions:

Artificial Intelligence In computer science, the field of AI research defines itself as the study of "intelligent agents": any device that perceives its environment and takes actions that maximize its chance of success at some goal. (Russell et al., 2010)

Machine Learning An agent is considered to learn "if it improves its performance on future tasks after making observations about the world." (Russell, 2016) The main feature of machine learning is the ability to learn and to adapt without being specifically programmed. By observing the world, aka reading data, an agent is able to recognize patterns in the data and draw conclusions.

Despite the ongoing discussions about whether ML is a subtype of AI, the definition that will applied in this work is that ML is perceived as a field of automated pattern recognition. Within manufacturing use-cases, ML is particularly useful given the vast amount of data that is produced on the lower levels of an automation system and which cannot be sensibly analyzed manually. Especially in the light of short product life cycles and high product variety, ML is able to play a major role in these transformation processes. In this context, ML is capable of learning from and adapting to changing environmental conditions "the system designer need not to foresee and provide solutions for all possible situations" (Alpaydin, 2010). The research field of ML is further distinguished into three main categories according to the literature, which are summarized under the terms *unsupervised learning*, *reinforcement learning* and *supervised learning* (Russell, 2016; Monostori,

2003). The main difference of these learning techniques is characterized by their feedback mechanisms (Monostori, 2003):

Reinforcement Learning Characterized by less feedback, since no proper action rather than only an evaluation of the action is given by the teacher.

Unsupervised Learning No evaluation of the action is provided, since there is no teacher.

Supervised Learning The correct response is provided by the teacher.

Each of these learning techniques is fulfilled by making use of concrete algorithms or toolboxes, of whom the most important or best-known are artificial neural networks, Support Vector Machine (SVM) or clustering approaches. These approaches are partly overlapping among the different learning techniques (see Figure 3.4).

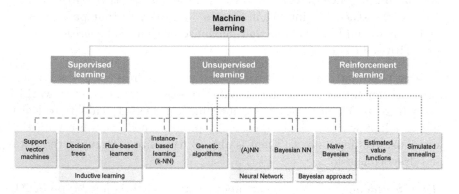

Figure 3.4: Machine Learning techniques and algorithms (Wuest et al., 2016)

The visualization in Figure 3.4 should not be understood as an exhaustive list of all possible ML technique, rather as a summary of the most commonly used methods. Each of method has specific properties that qualifies this method more for a distinctive family of tasks than the other methods. Some of the learning techniques can be further summarized under groups of learning mechanisms, e.g. for applications of neural networks.

3.7.2 Applications of Machine Learning in Manufacturing

As already mentioned in the previous section, an application of ML techniques is expected to be quite applicable in the area of manufacturing, especially due to its ability to automatically learn from data and to adopt to changing environmental conditions.

Unsupervised learning provides methods for finding patterns in data sets without a concrete labeling. It is an emerging field of research, especially in the context of Big Data applications. However, in many manufacturing systems, the data is assumed to contain at least some kind of label, e.g. a classification of states (Monostori, 2003). There might be some cases, in which expert feedback is not required, however, currently unsupervised learning techniques are not widely applied in manufacturing yet (Wuest et al., 2016).

Reinforcement learning is implemented into solutions of the manufacturing industry, however, only a few applications exist by today (Wuest et al., 2016). These applications include Günther et al. (2014), in which a hybrid approach of unsupervised and reinforcement learning is applied for a self-learning and self-improving laser welding system. Another application is examined in Lu (1990), wherein a reinforcement learning approach to support decisions during production ramp-up processes is proposed.

An examination of machine learning techniques in the context of manufacturing shows that supervised learning methods are mostly applied in the manufacturing industry, especially "due to the data-rich but knowledge-sparse nature of the problems. Supervised ML is applied in different domains of manufacturing, monitoring, and control being a very prominent one among them" (Wuest et al., 2016). In the literature, it is stated that supervised learning plays a highly dominant role in manufacturing as most applications are able to provide labeled data (Alpaydin, 2010; Lu, 1990; Harding et al., 2006). One important technique of supervised learning applications in manufacturing are SVM. Applications of SVM have been applied to various domains of industrial manufacturing, of whom the most important are:

- Tools for machine condition monitoring and fault detection in manufacturing have been examined by Azadeh et al. (2013); Salahshoor et al. (2010) and Zhong et al. (2010).

- Tool wear classification for machine tools in the manufacturing industry were examined by Sun et al. (2004).

- Quality monitoring is covered by Ribeiro (2005).

- Image recognition in terms of characters/faces was examined by Salahshoor et al. (2010), wheres image processing of damaged products is covered in Çaydaş and Ekici (2012).

- Time series and demand forecasting for the manufacturing industry has been examined by Salahshoor et al. (2010) and Guo et al. (2008).

- Finally, job shop scheduling and its various application in production optimization is covered in Baidya and Ghosh (2015).

The applications of these research fields cover a range from the machine level, i.e. machine condition monitoring, to process design, i.e. job shop scheduling, and production planning, i.e. demand forecasting. For this work, especially the scenarios that involve tool or machine conditions monitoring, fault diagnosis and fault detection are in focus. These fields of manufacturing, which can be summarized under the term of *predictive maintenance*, will be further examined in chapter 7 that concentrates on the learning of smart agents on the factory floor.

3.8 Information Modeling for Intelligent Automation

In the last section, the role of distributed intelligence on the shop floor, e.g. by means of low-level logic or through sophisticated machine learning algorithms has been described in detail. Despite the role of these techniques in the role intelligent automation, suitable information modeling approaches in order to bring those smart agents together with manufacturing planning and execution, are also of vital importance for the manufacturing process.

Existing approach target for example an interconnection of smart agents in the field with intelligent entities within adaptive MES (Ulewicz et al., 2012). These attempts try to combine the advantages that have been obtained by embedding smart agents either in the shop floor or within higher information systems. Architectures that focus on such recombination aim at an information and communication infrastructure that is established between system architectures of different time scales and runtime platforms (field devices and MES) (Ulewicz et al., 2012). These efforts provide an important contribution to the problem of incorporating loosely-coupled (service-oriented) and tightly-coupled (field buses) systems as described in the problem fundamentals of this work.

One important target of bringing together such systems is connected to a resolution of the traditional layers of the automation pyramid. Although a complete revocation of this hierarchical model cannot be reached entirely in the near future, different parties in the field control area discussed the reduction of levels in the automation pyramid and concluded that the structure of the pyramid will decrease onto two levels (Vogel-Heuser et al., 2009).

The authors of these approaches conclude that an integration of loosely-coupled and tightly-coupled system is only realizable in a generic manner, if interfaces with comprehensive specification of syntax and semantics can be established (Vogel-Heuser et al., 2009) on all levels of the manufacturing automation. The semantic specifications of these solutions are especially important to realize an intelligent transformation of information from low-level devices into a form that is readable by high-level systems.

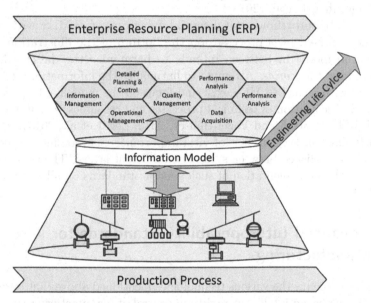

Figure 3.5: Diabolo as global information architecture according to Vogel-Heuser et al. (2009)

Until today, the integration of information from the field is mostly realized by proprietary solutions and tool-based information exchange mechanisms, "resulting in exhaustive programming of so called glue code and mostly unsatisfying solutions, especially if iterative development takes place. To achieve a real improvement in standardization [it is] required to take into account the different dimensions of information integration" (Vogel-Heuser et al., 2009). There is no common data basis or specification for the harmonization of different software solutions and any changes of existing software is connected to tremendous efforts (Thiel et al., 2010). Thus, the vertical integration from the low-level production layers with their embedded systems into the MES and ERP layers of an enterprise is one of the crucial challenges for the evolution of today's production systems to CPS (Vogel-Heuser et al., 2013).

The considerations and challenges result in the development of new topologies for automated production systems, e.g. as shown by the introduction of the double cone or Distributed Architecture to Bolster Lifecycle Optimization (DIABOLO) (Vogel-Heuser et al., 2009). The general idea and the resulting architecture of such an approach is shown in Figure 3.5 (adapted from Bauernhansl et al. (2014)).

The architecture shows the essential functions of a MES within the top cone and field level processes on the bottom of the illustration. The information model in between intends to harmonize the data exchange between these *two worlds*. Attempts to harmonize the information exchange especially with regard to MES exist and focus on the modeling language for a detailed specification of technical processes and resources, e.g. in terms of the MES-ML (Witsch and Vogel-Heuser, 2012; Legat et al., 2014).

Concluding this section, it is of vital importance that common standards need to be established between systems on different layers. The next section deals with the implementation of such generic standards on all layers of the manufacturing.

3.9 Advanced Interoperability Standards for Manufacturing

Taking into account the various technologies mentioned above that intend to enable means for an intelligent real-time control of manufacturing processes, there is still a lack of integrated technologies that provide a link between tightly-coupled systems and the requirements of open, dynamic environments. On the one hand, the conditions for real-time control and reactive automation require a support of grown manufacturing environments with their bus systems and Industrial Ethernet environments, on the other hand, dynamic and autonomous system environments, e.g. for the realization of multi-agent systems in production are demanded.

Approaches that attempt to tackle these challenges that are connected to the desired holistic interoperability in automation systems need to support legacy low-level systems as well as high-level planning operations by means of decentralized decision-making and distributed intelligent approaches. The most promising way to reach such kind of interoperability is to introduce standardized communication solutions that are capable of both integrated information exchange with low-level systems as well as share common ground in terms of semantics and information of high-level management and planning applications.

The solutions that had been elaborated in the previous sections focus more on specific aspects of this aimed interoperability than on a holistic approach:

- Bus systems and fieldbus protocols are designed from the interface point-of-view, i.e. the interoperability that is targeted through these systems focus on the actual machine interface that are used for the communication between machines, sensors, embedded devices or PLC. Thus, bus and similar protocols look more specifically at the low-level interface protocol, i.e. the "language" used by certain de facto standards that have been evolved within factories throughout the last decades.

- Industrial Ethernet and novel solutions that are represented by so-called next-generation fieldbus protocols as described in section 3.4.2 (e.g. PROFINET I/O or Sercos III [Bosch]) intend more on a combination of Internet-based applications (e.g. IoT) with manufacturing networks. These approaches intend to provide more generic interoperability approaches by means of service-oriented architectures, e.g. by making use of an ESB. Despite the fact that these solutions enable some sort of connectivity from the lower levels to high-level planning and control systems, the interfaces to the upper systems are rather hard-coded for specific proprietary applications than enabled through a generic interoperability.

- The real-time Industrial Ethernet solutions (TSN) as well as Publish-Subscribe architectures such as described in 3.4.4 and 3.4.3 go one step ahead by aiming at more flexible service-oriented architectures. By introducing the concept of communication channels and application-specific topics, e.g. for publish subscribe approaches like MQTT and DDS, interfaces between tightly coupled systems and high-level applications can be constructed in generic ways similar to REST definitions for web services. The flexibility of such implementations has been approved in many applications, e.g. in robotics or motion control use-cases that depend on information exchange in combination with the control real-time systems. Despite the flexibility and easy to implement adaptability of these approaches, they do not offer ways to enable semantic scalability in terms of information modeling or the usage of standardized ontologies. Publish subscribe approaches might be constructed to work well with the semantics of low-level automation systems, however there is no consistent *language* for the information exchange between the lowest levels of the automation and high-level management applications.

An attempt to incorporate the functionalities of the aforementioned standards together with the flexibility of service-oriented architectures of publish subscribe infrastructures is the meta standard OPC UA. Through the integration of the classical OPC standard that has been evolving throughout the last decades, low-level functionalities and connectivity with field level applications can be accomplished. Besides the support of flexible transport mechanisms, the OPC UA standard also provides dedicated information modeling approaches in terms of representing production data in a consistent semantic context. The scalability of these semantics might reach from the lowest levels of the automation pyramid up to the top-level planning systems like ERP, incorporating information from all layers in between, such as represented thtough SCADA, MES or PLM systems.

This section examines the functionality and practical usage of the OPC UA standard in detail. Section 3.9.1 focuses on the emerging of the OPC standard, its basic characteristics and its implementation into the industrial reality as a de facto standard for serving interoperability between low-level and diagnostics systems. The OPC UA standard is accordingly examined in section 3.9.2 and concentrates of the various aspects of this next-generation interface standard enabling integration of systems from all vertical layers by means of information modeling. After an introduction of the basic concepts, further application domains of OPC UA are targeted in the following section, focusing on the basic system architecture and service functionalities (section 3.9.3), the address space model and information modeling capabilities (section 3.9.4), security (section 3.9.5), web-services and data-driven approaches (section 3.9.6), real-time approaches through OPC UA in combination with TSN and publish subscribe approaches (section 3.9.7), enhanced information model extension through OPC UA companion standards (section 3.9.8) and finally existing approaches to integrate distributed intelligence concepts in terms of a realization through OPC UA (section 3.9.9).

3.9.1 OPC – The Interface Standard for Generic Shop Floor Interoperability

Throughout the last decades, especially in the 1980s and 1990s, the number of automation devices used in industrial manufacturing grew in an extensive amount. The commissioning efforts with regard to the configuration of automated processes based on PLC and similar automation devices was comparatively high, as most of the implementation work had to be performed manually (Rinaldi, 2013). In most of the applications, thousands of lines of ladder logic tailored to the usage in specific devices had to be implemented specifically for each application. The implementation of device-specific drivers was connected to tremendous work efforts that had to be renewed for each additional devices.

In order to limit the efforts for the commissioning and configuration of new devices in automated production systems, major automation companies got together in the beginning of the 1990s to derive a standard that offers generic access to the information of automation devices. In 1994–1995 these companies created the OPC 1.0 standard together with the establishment of the OPC Foundation. The goal of OPC was simply to provide a generic interface to arbitrary devices that could be implemented without having to know about the internal structure of that device. Accordingly, the OPC interface standard intended to carry out an information modeling standard that would allow manufacturers of automation devices to integrate a device-specific driver into their device. This driver would allow any application that *speaks* OPC to access the information of the device seamlessly (Rinaldi, 2013). Due to the high prevalence of Microsoft Windows based devices within industrial environments, the OPC communication standard was carried out based on the Component Object Model (COM)/Distributed Component Object Model (DCOM) technology (Brown and Kindel, 1998).

The COM/DCOM interface technology is based on the underlying Object Linking and Embedding (OLE) technology. In order to express the extension of OLE technologies in terms of industrial applications, the initial OPC was an abbreviation for OLE for Process Control. However, due to the fact that newer applications of OPC do not rely on the OLE standard anymore, the old abbreviation has become deprecated. Thus, over the years, the term OPC in itself has had many terms tied over it (Hannelius et al., 2008). At present, the interpretations of the OPC term has changed to "Open Platform Communications" respectively "Openness, Productivity, Collaboration". Newer interpretations do not even refer the OPC term to a concrete abbreviation, thus OPC stands representatively as a name on its own for a series of interoperability standards (Lange et al., 2014).

Based on the generic interoperability of the COM standard within the Microsoft Windows domain, every application that is runnable on the Windows operating system is by default able to exchange information with other application. The extension of this concept towards drivers for automation devices provided the easiest approach to enable a generic interoperability between hardware devices and Windows-based hardware controllers and supervisory systems in the manufacturing environment. The fist OPC standard that was carried out in the middle of the 1990s finally provided an API that takes into account the COM standard and provides an interface for device-specific drivers accessing the information of the actual hardware device (see Figure 3.6).

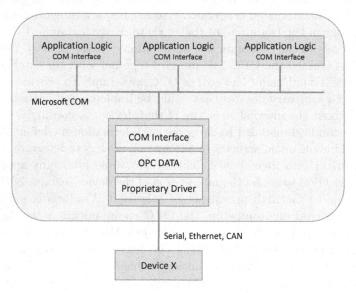

Figure 3.6: COM interface to shop floor devices

Technically, the link of modern protocols like OPC Data Access or OPC Unified Architecture is realized by gateways that formed the basis for traditional field level protocols such as Modbus or 3964R. The 3964R protocol is a point-to-point serial protocol that enables communication between two Programmable Logic Controllers (PLC) (e.g. commonly used in Siemens S5 devices) and functions according to the master/master paradigm, i.e. the

PLCs are able to configure and manipulate each other's memory components autonomously. This "last mile" interface to the low-level technical layer (Second layer in OSI model – Link Layer) can be utilized by modern interface standards to reconfigure the technical implementation of production control logic and hence reorganize manufacturing processes in real-time.

As long as developers of hardware devices provide a proprietary driver that supports underlying bus protocols as well as the information representation paradigms of OPC, any other application above the shop floor is able to access the information of such devices through the COM interface in a generic fashion. The relocation of driver development from the software side directly to the hardware vendors was a revolutionary approach and induced tremendous savings in terms of work efforts when connecting a new device on the shop floor. The OPC standard provided a hardware driver support through which the development of application-specific became entirely obsolete. Every hardware component supporting OPC could be connected to arbitrary software applications by simple plugging in the device to an OPC conform interface – just like an USB stick.

The initial set of standards published by the OPC Foundation consists of the OPC Data Access (DA) specification, the Alarms & Events (A&E) and Historical Data Access (HDA) specification (Leitner and Mahnke, 2006). The DA specification describes the access to current process data, the A&E describes an interface for event-based information exchange including acknowledgment of process alarms and the HDA specification provides functionalities to access historical/archived data (Mahnke et al., 2009).

OPC makes use of a client-server approach to realize information exchange. An OPC server, which is usually within or in the near range of automation devices, encapsulates the source of information like a device and makes this information available via the OPC interface. Any OPC client that operate in the same network infrastructure is able to connect to the OPC DA server and can access or consume the offered data (Mahnke et al., 2009). Applications that make use of the OPC interface can incorporate both client and server instances. A typically example of OPC based interconnected applications is shown in Figure 3.7.

On the bottom of Figure 3.7 the data sources of an automation systems by means of PLC and DCS are depicted. These systems that are typically connected to sensors, machines and other manufacturing resources, connect the hardware in the shop floor through bus systems and other proprietary interfaces. By applying the OPC model definitions vendor specific interfaces/drivers to an OPC DA server can be carried out in order to represent

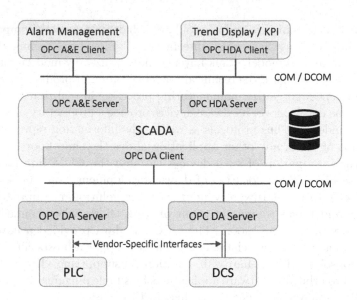

Figure 3.7: Typical configuration for vertical interoperability through OPC
servers and clients (adapted from Mahnke et al. (2009))

the device information through the OPC DA server instance. These interfaces are implemented in the form of a hardware specific driver that is embedded directly into the device or into a connector module that serves interoperability between the device and the according OPC DA server.

After the production data is made available by the OPC DA server instances, various client applications are able to access field level information directly through these servers by means of the generic COM protocol. Such OPC DA client might be embedded within a SCADA application in order to provide real-time information from the field directly to overall supervisory systems. Through the implementation of data persistence solutions, e.g. in terms of an embedded database that SCADA is able to further propagate the information to other systems. To achieve this interoperability, the SCADA system implements two additional server instances, i.e. one OPC server for alarms and events and another OPC server for historical data access. The according counterpart clients within the desired target applications are able to read from these servers through the COM/DCOM interface and provide the interesting information to the various applications of high-level planning and control systems.

The main advantage of the described approach consists in the reduction of specification work that was initially connected to the development of specific APIs regarding special needs and requirements of the underlying systems. By making use of the COM/DCOM interface their is no need to define specific network protocols or mechanism for interprocess communication. The COM/DCOM interface provides a transparent mechanism for any client to call methods on COM objects that are located in a server and which are running in the same process, in another process or even on another network node. The usage of this technology facilitates the development of applications, because the underlying transport protocol is already available on all PC-based Windows operating systems. Accordingly, the specification and configuration efforts for new products and their time-to-market can be reduced significantly. The prevalence of the used standard in existing automation environments was one major factor for the success of OPC (Mahnke et al., 2009) and its evolving to a de facto standard within industrial manufacturing.

3.9.2 OPC UA – The Interface Standard for Integrated Communication and Information Modeling

The major success factor of OPC DA, which is nowadays also referred to as OPC Classic, consisted in the adoption of an established Microsoft Windows interface standard. However, due to the emerging prevalence of embedded devices, IoT applications and an increasing implementation of UNIX-based systems, e.g. for motion control and robotics applications, the limitation of the communication to COM/DCOM is rather impeding for many applications in modern automation environments. In order to tackle these interoperability issues, further extensions of OPC were proposed. The first attempt of the OPC Foundation to maintain the successful features of OPC while using a vendor and platform neutral communication infrastructure consisted in the OPC XML-DA specification. The idea behind the underlying concepts is to introduce Web Service solutions for common OPC functionalities based on XML messages. However, due to the poor interoperability of XML based Web Services as well as significant performance losses in comparison to native OPC DA an extensive adoption of the OPC XML DA standard in practical applications did not succeed (Mahnke et al., 2009).

The described limitations in terms of generic interoperability with vendor independent operating systems as well as some other drawbacks regarding the usage of OPC Classic finally led to the necessity of developing a new OPC standard. The most decisive weaknesses of OPC Classic that need to be overcome in terms of a next-generation OPC standard are summarized in the following (Hannelius et al., 2009):

- Microsoft introduced the .NET Framework in 2002 which is based on Web Services instead of COM and DCOM. The COM and DCOM technology is still integrated into actual versions of Microsoft's operation system, but it is the first step towards a new base technology.

- Every OPC Classic specification uses its own data model. OPC vendors requested a single set of services to handle data models for different purposes like Data Access, Alarms & Events, Historical Data Access, etc.

- OPC users are in need of an easy, fast and secure access to OPC servers in complex networking environments. However, using the COM/DCOM standard, it is especially complicated to enable a communication through firewalls.

- OPC vendors called for an OPC implementation that interoperates independently of Microsoft Windows operating systems in order to stay competitive. For example, an implementation of OPC components on power saving ARM based hardware like the well-known Raspberry Pi is not possible through OPC Classic implementations.

The development of a new specification taking into account the aforementioned limitations and drawbacks of classical OPC led to the specification of a new standard – OPC Unified Architecture. The design goals of the OPC UA specification consist in the integration of all OPC Classic functionalities into one platform independent, secured and extensible framework. Unlike OPC Classic, OPC UA does not depend on any technologies of proprietary Operating System (OS), thus its lifetime is not limited to the maintenance of any underlying technology.

By making use of a SOA for realizing the communication between entities, the modeling of the communication model can be performed independently from proprietary base technologies. Thus, OPC UA focuses on widely accepted standards and vendor independent solutions, e.g. based on web service technologies which are standardized by the World Wide Web Consortium (W3C). The core technology of the underlying information exchange techniques is XML (Booth et al., 2004), however OPC UA also enable data exchange between two endpoints by means of a binary encoded protocol (Hannelius et al., 2008).

OPC based on DCOM was "designed for straight forward integration into function block based basic automation information and thus needed no complex data model or address space" (Hannelius et al., 2008). Another major advantage of OPC UA in comparison to OPC Classic is its information modeling capabilities that are inevitable for the integration into complex applications. Especially these capabilities will be of major focus in the course of this work.

One of the central cornerstones that makes OPC UA to a unique standard among all the other specifications is its concept for separation of concerns (see Figure 3.8)

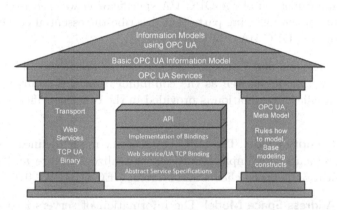

Figure 3.8: The two pillars of OPC Unified Architecture (Mahnke et al., 2009)

The two pillars form the basic foundation of OPC UA. The entire standards is build upon these two pillars as cornerstones for all OPC UA based communication.

The left pillar represents the transport capabilities of OPC UA that can be adapted to the requirements of each application in a flexible way, either through low-level protocols based on TCP or web services that allow for a communication in scalable SOA. Regarding available protocols, the information exchange can either be realized through a highly performant binary protocol or based on a scalable XML-based protocol stack. The right pillar represents the OPC UA meta model which is responsible for the information modeling abilities of OPC UA. This meta model provides basic modeling rules for the derivation of OPC UA information models, which could be compared with fundamental concepts about object-oriented modeling. The characteristic feature of OPC UA specifically consists in the separation of these pillars. Thus, the information modeling part of an OPC UA application can be carried out completely independent from the concepts that realize the information transport in an OPC UA based network. This separation of protocols and semantics makes OPC UA to a unique standard.

On the top of the pillars OPC UA services are located that enable communicating entities to use a combination of the transport and information modeling capabilities of OPC UA. The basic OPC UA information model as well as additional information models provide concrete implementations of models that have been derived by means of the meta model definitions.

The standardization of the OPC UA specification was released in 2007 and consists of the following parts that describe the essential cornerstones of implementing OPC UA solutions:

Part 1 – Overview and Concepts Basic concepts about the address space and message model as well as the communication and security models. A more detailed examination is provided in the other specification parts. (IEC 62541-1, 2015)

Part 2 – Security Model The OPC UA security model defines the security requirements and implementations to realize a secure and reliable communication between clients and servers. (IEC 62541-2, 2015)

Part 3 – Address Space Model The information of servers is organized with a data structure that is referred to as the *AddressSpace*. The underlying address space model consists of nodes that are interconnected through *References*. A node is derived from one of eight base node classes and can be specified in terms of *Attributes*. (IEC 62541-3, 2015)

Part 4 – Services Examines the services that are used to realize the communication between OPC UA servers and clients. Common services are characterized by the requirements for *discovery* capabilities or the realization of monitoring or subscriptions to node values. (IEC 62541-4, 2015)

Part 5 – Information Model The information model describes an extensible methodology to standardize nodes that can be added to a server's *AddressSpace*. (IEC 62541-5, 2015)

Part 6 – Mapping This part links the abstract services and information definitions of the previous parts to concrete technologies. (IEC 62541-6, 2015)

Part 7 – Profiles Profiles are subsets of OPC UA functionalities. In practical use-cases, OPC UA applications will only utilize a fraction of the possible functionalities. Thus, vendors developing OPC UA based applications are encouraged to make use of these profiles in order to define the functionalities of their frameworks. (IEC 62541-7, 2015)

Part 8 – Data Access Incorporates the functionalities of classical OPC in terms of the DA module. (IEC 62541-8, 2015)

Part 9 – Alarms and Conditions Incorporates the functionalities of classical OPC in terms of the A&E module. (IEC 62541-9, 2015)

Part 10 – Programs Specifies the development of programs that are located on an OPC UA server and can be executed and administered by clients. (IEC 62541-10, 2015)

Part 11 – Historical Access Incorporates the functionalities of classical OPC in terms of the HDA module. (IEC 62541-11, 2015)

Part 12 – Discovery Describes the *Discovery* service that can be used by clients in order to find servers and access information of the specific endpoints within an OPC UA based network architecture. (IEC 62541-12, 2015)

Part 13 – Aggregates With respect to the information modeling, the *Aggregates* specification mainly extends the capabilities of servers to support aggregates that can used for historical or current data. The main objective of the part is to describe aggregation functions for the various types of information that can be gathered in the server's *AddressSpace* (Mahnke et al., 2009). (IEC 62541-13, 2015)

The following sections give a quick introduction into the basic characteristics of OPC UA regarding their importance for later parts of this work. For a deeper introduction to all aspects of OPC UA, the interested reader might refer to IEC 62541 (2015), Lange et al. (2010), Mahnke et al. (2009) and Mahnke et al. (2011).

3.9.3 OPC UA System Architecture and Basic Services

According to IEC 62541 (2015) OPC UA was carried out to enable a robust propagation of data independent from its source. The major area of application for the standard is the manufacturing industry. OPC UA enabled software is designed to integrate information from all levels of a manufacturing company, from field devices over PLC until the front end of ERP and PLM systems.

OPC UA does not specify an explicit API to realize the information exchange, but only the message formats for data exchanged on the wire. The client as well as the server-side of an OPC UA network make use of a *communication stack* to encode and decode message requests and responses. One strength of OPC UA is that different communication stacks can work together seamlessly as long as they support the same technology mapping (Leitner and Mahnke, 2006).

The system concept of an OPC UA infrastructure is based on the server-client concept. A system can consist of an arbitrary number of servers and clients in terms of interacting entities. Every server is able to interact simultaneously with several clients, every client is able to be connected to multiple servers. Unlike OPC Classic, an entity within an OPC UA based network infrastructure is able to represent both, a client as well as a server instance. Possible interaction scenarios between servers and clients are visualized in Figure 3.9.

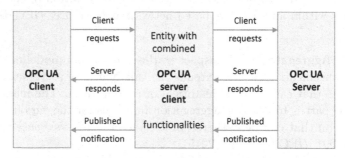

Figure 3.9: OPC UA system architecture based on clients and servers

The interaction between a server and a client is initiated by the client. If the client is not aware of the server's endpoint information the *discovery* service allows for a browsing of available servers including their endpoint information such as IP and port. If the endpoint information is already known, the client can establish a connection to the server via a *Session* request. If the server reacts positively to this request, the connection will be set up. As soon as a connection is established the client is able to request information from the server. Clients are also able to request a server to monitor its information for the client by subscribing to certain nodes. The server will accordingly publish the values of the subscribed nodes based on determined *triggers*, e.g. data change events or fixed publishing intervals.

For interacting with the server, a client application makes use of the OPC UA client API which is an internal interface that separates the client application from the OPC UA communication stack. The communication stack handles the message transfer between servers and clients including decoding, encoding, security encryption, by providing binary transport and protocol layers like TCP. For a detailed examination of the OPC UA communication stack the interested reader might refer to Leitner and Mahnke (2006), Mahnke et al. (2009) or IEC 62541 (2015).

3.9.4 AddressSpace Model and OPC UA based Information Modeling

The main component of an OPC UA server is its *AddressSpace*. The *AddressSpace* can be understood as an object-oriented structure of OPC UA nodes that are represented by the server. These nodes are used to model real objects from the manufacturing environment and their according functionality. Nodes can be used e.g. to represent sensors as well as their according measurement values and certain context information. *References* are used to organize these nodes within the *AddressSpace*. The basic setup of the OPC UA *AddressSpace* is shown in Figure 3.10.

The OPC UA root node represents the base entry point into the *AddressSpace* structure. The *Types* folder contains all node definitions in terms of existing or custom node types. The *Object* folder contains actual instances of previously defined types. The *Server* object is a self-representation of the OPC UA server within its own *AddressSpace* and is a default component in the *Objects* folder. The *Views* folder contains customized views that are able to limit *read* or *write access* to certain nodes within the *AddressSpace*.

The information about real world objects that is represented through the *AddressSpace* can be structured in terms of the OPC UA meta model. This meta model provides concrete definitions of node classes that are used to model entities for the mapping of manufacturing information. *AddressSpace* entities are modeled in terms of *objects, variables, methods* and relationships between entities (*references*). Each entity of the *AddressSpace* is modeled in

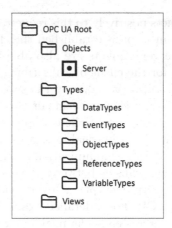

Figure 3.10: Basic structure of the OPC UA *AddressSpace*

the form of a *node*. *Objects* are defined as a set of *nodes* that are described by *attributes* and interconnected through *references*. A certain *node* is derived or initialized by using one of eight base node classes. These node classes are:

- Variable
- Method

- DataType
- ObjectType

- Object
- View

- VariableType
- ReferenceType

Every node within the *AddressSpace* belongs to a *Namespace*. An *AddressSpace* can support various *Namespaces*, which are identified by a Uniform Resource Identifier (URI). *Namespaces* are used to group certain sets of node definitions, e.g. based on the vendor of the underlying information model or for the separation of certain application domains and their according node definitions. The *NamespaceIndex* is an incremental number identifier for *Namespaces* starting at zero. For example, the basic *Namespace* of the OPC UA specification which is present in every OPC UA server instance is characterized by the *Namespace* URI http://opcfoundation.org/UA/ and the *NamespaceIndex* "0". Additional information models, e.g. for the definition of domain-specific nodes that are integrated into the same *AddressSpace* will be accordingly identified by their URI and *NamespaceIndices* 1, 2, et cetera.

Every node that is located in a certain *AddressSpace* needs to provide a unique identifier. The *NodeId* consists of the *NamespaceIndex* as well as an additional identifier for the node that can be an Integer or a String:

Listing 3.2: Different NodeId unique identifiers

```
NodeId with Integer Identifier: ns=1;i=28382
NodeId with String Identifier:  ns=1;s=CastingMachine.
  Temp.Sensor2
```

The first *NodeId* definition makes use of a unique Integer to identify the *node* within the *Namespace*. The usage of the "i" for signalizing that an Integer is being used for identification facilitates the parsing of *NodeIds*. The second line in Listing 3.2 might refer to the same node as the previous one, however using a String as *node* identifier. The String represents the object-oriented *path* according to the tree structure of the *AddressSpace*.

The concatenation of the *NamespaceIndex* and the according identifier makes the *node* identifiable throughout all OPC UA servers and *AddressSpaces* within the OPC UA network infrastructure. This concept is used to allow several OPC UA servers to represent the value of the same *node*, i.e. real world object, while maintaining the unique identification of that *node* throughout all servers.

The actual modeling of custom information models is realized by extending existing *node* definitions based on the *node classes* mentioned above. The according *node class* determines the attributes and references which have to be set during the initialization of a *node* within the *AddressSpace*. Attributes and references are elementary components of *node classes*. Attributes are defined by setting an attribute ID, a name, a description, a data type and a *mandatory* flag. The *mandatory* flag determines whether an attribute has to be set during the initialization of the according *node* or not. The data type of an attribute can be a primitive data type like Integer, Double, String, etc. or a custom (complex) data type defined in terms of a *Variable*. The attributes of a *node* can be accessed by clients in terms of *read* operations and other services defined in Part 4 of the OPC UA specification.

References are used to show certain relations between two *nodes* within the *AddressSpace*. Every *Reference* are derivatives of the basic *ReferenceType node*. There are different basic *References* that are well-defined by the OPC UA base information model. These *References* are classified in hierarchical and non-hierarchical ones. Hierarchical *References* are used to derive hierarchical structures between *nodes*, e.g. in terms of the *HasComponent Reference* which is used to compose objects based on multiple other nodes. An example for a non-hierarchical is the *HasTypeDefinition Reference* that is used to derive *nodes* based on an extension of other (parent) *nodes* in an object-oriented fashion, i.e. by using the concept inheritance.

A node that contains a *Reference* is referred to as *Source Node*. The node this *Reference* points to, is called *Target Node*. An example for a concrete *Reference* in terms of information modeling is the *HasComponent Reference* (see Figure 3.11).

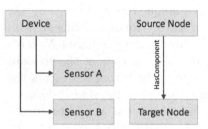

Figure 3.11: *HasComponent Reference*

```
<!-- The Device XML Structure -->
<UAObject NodeId="i=100" BrowseName=
    "Device">
<References>
  <Reference ReferenceType="
      HasComponent">
    i=300 <!-- Points to SensorA -->
  </Reference>
  <Reference ReferenceType="
      HasComponent">
    i=301 <!-- Points to SensorB -->
  </Reference>
</References>
</UAObject>
```

In the figure the *HasComponent Reference* is used to allocate two sensors to a device. *Sensor A* and *Sensor B* are two separate *nodes* representing physical sensors. The *HasComponent* references describes that they are part of the device in the form of a *composition*. The code on the right shows the same relation in terms of the XML representation of OPC UA information models.

The information modeling capabilities of OPC UA as shown in the example above can be used for the development of extensive and information models representing complex object structures and type definitions. The information model creation using this meta model approach will be of major importance in later parts of this work, in which the OPC UA based representation of intelligent entities by means of software agents and MAS are carried out.

Due to limited space, not all meta modeling capabilities of the OPC UA specification can be examined here. For further detailed instructions on the modeling methodologies of OPC UA the reader might refer to IEC 62541 (2015) and Mahnke et al. (2009).

3.9.5 Security Aspects of OPC UA for Reliable Information Exchange

Despite the enthusiasm that is connected to concepts like the factory of the future, in which everything is interconnected in terms of an IoT environment, some critical discussions concerning the security of interoperating manufacturing units have emerged. The OPC UA standard does not only serve interoperability among various layers of the manufacturing hierarchy, the specification also takes into account modern concepts for encryption and secured information transport technologies. The OPC UA security specification relies on existing state-of-the-art security standards and implements all important security features into the OPC UA based communication. The security objectives of the standard are precisely:

- Authentication – Methods to determine the identity of an entity

- Authorization – Granting access to system resources and functionalities

- Confidentiality – Protection of information from unauthorized access

- Integrity – Production of messages from subsequent corruption

- Auditability – Regular system checks to determine its safety status

- Availability – Resource disposability of the system

OPC UA provides a number of protocol option and additional features to enable the security requirements stated above. As the actual information exchange through OPC UA can be performed in terms Web Services or through the light-weight UA Binary protocol, security modes for both transport mechanisms are proposed. The *Web Service Secure Conversation (WS-SC)* provides security means for Web Service based message exchange. The UA Secure Conversation (UA-SC) guarantees secure data transfer when communicating through the binary encoded protocol mode. The transport security for all transport modes is enabled by Secure Sockets Layer (SSL)/Transport Layer Security (TLS) (Hausmann et al., 2015).

The authentication of OPC UA clients and servers is performed by means of X.509 certificates. The establishment of a secure channel with integrated key exchange techniques is realized by means of the Rivest-Shamir-Adleman cryptosystem for public-key encryption (RSA). The protection of payload information is achieved by means of symmetrical algorithms, namely Advanced Encryption Standard (AES) and Keyed-Hash Message

Authentication Code (HMAC). The most reliable security guidelines that can be activated for the communication through OPC UA are referred to as *Basic256SHA256*, which a *symmetrical signature (MAC)* of *HMAC SHA-256*, a *symmetrical encryption* of *AES-256 CBC*, *asymmetric signatur* of *RSA SHA-256*, *asymmetric encryption* of *RSA OAEP* and an *asymmetric key length* of *2048-4096 Bit* (Hausmann et al., 2015).

As stated by the German Federal Office of Security in Information Technology (Bundesamt für Sicherheit in der Informationstechnik – BSI) that security level of OPC UA can be regarded as high if the *SignAndEncrypt* security mode is configured in the server (Wichmann, 2016).

3.9.6 Service-Oriented and Event-Driven Approaches through OPC UA

The flexible architecture of OPC UA in terms of arbitrary client server infrastructures offers means to design event-driven interactions and process mediation. Applications that are characterized by such event-driven infrastructures might consist for example in ESB or other architectures with rich service infrastructures.

In terms of OPC UA, applications for complex event processing are often rolled out by means of coexisting OPC DA (OPC Classic) and OPC UA services. Hereby, the data access entities are placed at the bottom of the chip level enabling direct interactions with field level devices (Izaguirre et al., 2011). The OPC UA instances are used to incorporate the OPC UA legacy systems and mediate certain information to the upper levels of the manufacturing automation system and management levels. In contrast to these limited interoperability capabilities, the transition from OPC Classic to OPC UA on all levels opens up to standardize the processing of events and their according data basis in a consistent way. In this context, "DPWS and OPC UA are becoming nowadays the preferred options to provide on a device level, service-oriented solutions capable to extend with an Event Driven Architecture into manufacturing systems" (Izaguirre et al., 2011).

The capabilities of DPWS in combination with OPC UA in terms of incorporating devices on the shop floor have already been examined (Candido et al., 2010). The researchers concluded that consistent interoperability cannot be achieved in terms of one single technology, thus the combination of interface standards like OPC UA and semantic service definitions in terms of DPWS are needed. One example are the capabilities of DPWS to create customized discoverable services and events as well as providing plug-

and-play features and mapping all internal operations on devices (Izaguirre et al., 2011). In combination with the information modeling capabilities of OPC UA and additional mediators and gateways, it is possible to integrate most of the shop floor equipment using these technologies (Karnouskos et al., 2009).

The realization of event-driven approaches and architectures through OPC UA is achieved based on subscriptions, custom triggers, data change events, alarms and condition monitoring and similar approaches. The asynchronous communication model of OPC UA enabling arbitrary interfaces between clients and servers accordingly enables an establishment of complex events depending on explicit rulesets. A successful recombination of web service profiles and the aforementioned eventing mechanisms of OPC UA can be used in terms of a Complex Event Processor (CEP) (Luckham, 2006) that is part of the middleware and is accordingly able to correlate and aggregate the events coming from OPC UA (Izaguirre et al., 2011).

Existing solutions address possible technology merging strategies of OPC UA and DPWS (Minor, 2011):

- Candido et al. (2010) propose the SOCRADES framework with the purpose of a generic integration on device-level. Their solutions in terms of serving interoperability between the two technologies consists in a two-way mapping between device data and the object model of OPC UA by means of a low-resource version of a combined protocol (Izaguirre et al., 2011).

- Bony et al. (2011) demonstrate an approach that makes use of an OPC UA *node manager* that manages devices and applications through the OPC UA *AddressSpace*. The approach furthermore proposes shared libraries between an *ANSI C Stack* for the mapping of the OPC UA Binary protocol and a *DPWS Stack* to access the *nodes* within the *AddressSpace* either through *OPC UA over WS* or *DPWS WS*.

A possible implementation of these approaches could be realized by modeling entities of the available DPWS services in terms of custom *nodes* in the *AddressSpace* and access these entities via OPC UA *Method* calls. The arising solution can be especially beneficial, if the available high-level systems are not capable of suitable OPC UA interfaces or if other applications are integrated in terms of an ESB infrastructure. In the current, a similar approach to these web service integration strategies is proposed in the use-case section by incorporating a MQTT connector for high-level systems into the service architecture of OPC UA.

3.9.7 Interoperability with Tightly Coupled Systems by Means of OPC UA and Time-Sensitive Networking

The configuration and setup of connections is an important factor for the reliability and security of an information exchange technology. Additional features like automatic reconnection or caching of messages might be of importance in terms of a reliable data exchange.

The communication of OPC UA is principally based on established connections by making use of sessions. The establishment of such connection is realized by means of a secure channel that ensures the aforementioned security standards and reliability features. The connection set-up is performed by means of the discovery service (see Figure 3.12).

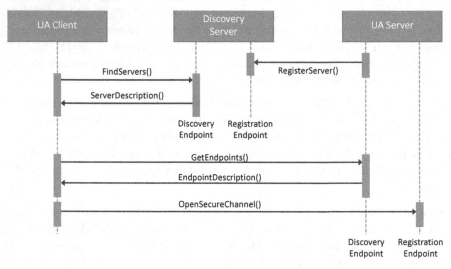

Figure 3.12: Discovery service of OPC UA to establish a secure channel based connection

In the first step of this process according to IEC 62541-4 (2015) the OPC UA servers register themselves at the *Registration Endpoint* of the *Discovery Server* by means of the `RegisterServer()` function. Consequently, the *Discovery Server* is capable of managing the connected servers in terms of providing their endpoint descriptions, identifiers and other additional information about the connected servers. After the registration of several servers, the `FindServers()` method will return an array of server information such as port or IP address through the `ServiceDescription()` method. Using this information the client is able to directly access the OPC UA server and requesting detailed endpoint descriptions by means of the `GetEndpoints()` method. The OPC UA server provides an `EndpointDescription()` and the client is able to establish a connection to the server by establishing a secured communication channel through the `OpenSecureChannel()` method according to IEC 62541-1 (2015). After this secure channel has been setup the client and the server are capable of communicating each other with respect to the available services (Damm, 2014).

The need to establish a secure channel prior to any communication between clients and servers might result in performance issues, especially in cases that require quick feedback times or instantaneous reactions. In order to compensate possible performance issues connected with these connection requirements the OPC UA standard provides several mechanisms to increase the performance of data transfer. In that way it is possible to track specific items in terms of data changes and asynchronously request information from the according data nodes. Such kind of behavior can be realized by means of several functionalities (Pessemier et al., 2011):

- OPC UA specify an upper limit or a fixed interval for incoming data that fits to its resources.

- Determination of a polling interval for data from certain *nodes*. The according values will be received in a certain time range near the expected target time. Through the specification of the polling interval prior to data exchange bandwidth overhead can be minimized during the actual data transfer.

- Filter functions to deal with noisy or fuzzy data.

- OPC UA servers are capable of setting a flag to data that has been already sent to the client. In case of recurring requests the server is able to resample this data.

Besides these *native* functionalities of OPC UA to increase performance in data transfer, the OPC Foundation is currently working on an extension of the OPC UA standard in terms of a Pub-Sub module (Damm, 2015). According to several resources (Unified Automation, 2015; Auberg and Stöger, 2016; B&R Automation, 2015) the according standardization is capable of enabling TSN for OPC UA bases communication. The realization of such approaches is performed by adapting the basic communication paradigms of OPC UA from traditional point-to-point communication in terms of request-response patterns to new paradigms realized by means of multi point connections via UDP. According to Auberg and Stöger (2016) the integration of TSN features into the OPC UA specification fits the fundamental concepts of OPC UA as the standard "was never bound to a fixed network topology, but was always capable of adapting [its structure] in terms of the request of the networking [requirements]". The flexibility of integrating OPC UA functionalities with *Publish-Subscribe* patterns becomes obvious in terms of the targeted information model of the new standard (see Figure 3.13).

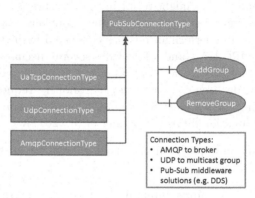

Figure 3.13: OPC UA Pub-Sub information model

According to the upcoming specification the user is able to choose between three different connection types. The TCP based *UaTcpConnectionType* represents the traditional OPC UA server client connection. The *UdpConnectionType* however is characterized by a connection method that is based on *secure multicast*. Finally the *AmqpConnectionType* represents the standard that is of main interest for the *OPC UA Pub-Sub Working Group*. This connection type intends to work in real-time critical environments, in which conventional OPC UA based data exchange is not suitable due to insufficient performance. The usage of AMQP is motivated by a continued cooperation of the OPC Foundation with Microsoft, who is using the message queue standard within its service bus architecture for the Microsoft Azure cloud (Microsoft, 2017). As the final Pub-Sub specification of OPC UA will be most likely based on the AMQP technology, the performance of deterministic OPC UA in terms of publish subscribe architectures will highly depend on the technical abilities of AMQP. However, the implementation of AMQP has only been performed in large scale systems yet, such as banking networks and similar applications (Spinnarke, 2016).

An important cornerstone with regard to the implementation of OPC UA consists in its integrability with standard OPC UA. Similar to the shift classical OPC to OPC UA, the OPC Foundation aims at realizing the transition from non-deterministic to deterministic data transfer smoothly and without compromising existing OPC UA implementations and use-cases. Thus, the upcoming Pub-Sub standard of OPC UA will be usable in combination and simultaneously to the OPC UA session based communication (see Figure 3.14). The target objective of this approach is to apply OPC UA for TSN only on critical points, where real-time communication is inevitable.

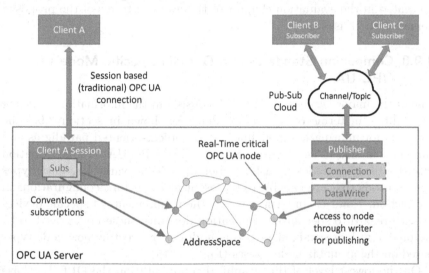

Figure 3.14: Simultaneous utilization of OPC UA publish subscribe models and session based information exchange, according to Damm (2015)

On the left side of the figure, conventional information exchange is realized by means of a client connection to a server through a session based secure channel. *Client A* connects to the server and accordingly subscribes to two different nodes within the *AddressSpace*. On the right side of Figure 3.14 the publish-subscribe model is visualized. The *DataWriter* connects to real-time critical nodes and transfers the according data values to a *Publisher*. This *Publisher* accordingly pushes the interesting data to respective channels within a *publish subscribe cloud*. Various clients subscribing to the according topics are able to read the published values within a small time frame.

Another promising approach to enable real-time capabilities through OPC UA consists in a combination of OPC UA with other broker-based or publish subscribe standards such as MQTT or DDS. In terms of DDS promising collaborations are already being carried out in terms of a "OPC UA/DDS Gateway" respectively an "OPC UA DDS Profile" for incorporating OPC UA communication into a DDS network (Real-Time Innovations, 2016). These initiatives represent a promising approach towards a light-weight and easy to integrate way to enable real-time capabilities for OPC UA. A collaborative usage of both, OPC UA and MQTT will be presented in the evaluation chapter of this work in terms of the proposed demonstration use-case.

3.9.8 Companion Standards and Domain-Specific Models for OPC UA

One of the major features of OPC UA consists in the information modeling capabilities in terms of its metamodel. As shown in section 3.9.4 the concepts for information modeling rely on object-oriented paradigms and can be expanded to an extensive amount. The OPC UA basic information model already includes a vast number of objects, variables, data types and methods that are capable of representing important configurations in production. However, in terms to meet the requirements of specific devices or with regard to specialized application domains, these nodes need to be adapted, e.g. by recombining existing nodes or by creating new node types based on the available node classes (Figure 3.15).

On the lowest level of this graphical representation the OPC UA base model is located that consists of the built-in nodes mentioned above. On top of this base model the incorporation of classical OPC standards is indicated by referring to the according specifications. Especially the *Alarms & Events* as well as the *Historical Access* are still representing vital and widely-used parts within OPC UA applications, the *Data Access* module is accountable for the integration of legacy OPC applications into modern OPC UA based environments.

Above the OPC UA basic and legacy modules, additional information models can be implemented in order to map domain-specific knowledge about certain resources (e.g. field level devices) or application domains (e.g. specialized automation technologies) on the object-oriented model

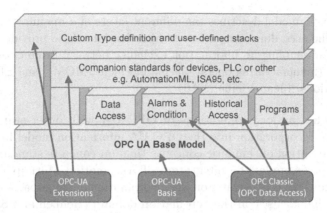

Figure 3.15: Generic extensibility of the OPC UA information model

of OPC UA. This way, specialized and process-related knowledge can be incorporated into the communication models seamlessly and without making use of proprietary formats or information representation methodologies. These so-called companion standards cover domains of major interest and are often standardized through cooperation or joint projects of the OPC Foundation with the according domain experts. Some of the most noteworthy companion specifications are mentioned in the following:

- The *PLCOpen* companion specification standardizes an information model for the mapping of function blocks in accordance to the DIN EN 61131-3 (2014) specification for programmable logic controllers.

- The Analyzer Device Integration (ADI) specification focuses on specialized information models for analyzer devices, such as sensors (ABB Group, 2010).

- The Field Device Integration (FDI) and Field Device Technology (FDT) companion standards (FDI Cooperation, 2011) focus more on field devices and their special information technological backgrounds, e.g. for the mapping of sensors like flow control or actuators such as valves.

- As an application of OPC UA is not limited to the manufacturing domain, standards of other related domains can also be incorporated with the metamodel of OPC UA. One example is the integration of the Building Automation and Control Networks (BACnet) standard for building automation and control systems according to DIN 16484-5 (2011).

- The ISA-95 model describes the different layers of a manufacturing company. The according companion specification intends to provide semantic means for enabling an information exchange between production related systems (manufacturing operations management (MOM)) and ERP/MES applications. (Jean Vieille, 2010)

Some other major cooperations of OPC UA with other standards in terms of an incorporation into the OPC UA information model focus more on the integration of semantic concepts than on concrete standardization efforts. For example, the integration of Automation Markup Language (AutomationML) concentrates on the incorporation of namespaces of a meta language that focuses on the automation domain (Henßen and Schleipen, 2014).

A meta layer between OPC UA and business applications is provided by the MTConnect standard that intends to realize a consistent information exchange between manufacturing equipment and the business layers by means of structured XML instead of proprietary formats (MTConnect Institute, 2013). Through the incorporation of the standard by means of the OPC UA meta model, the uniform nomenclature for production equipment can be integrated into the object-oriented mapping of OPC UA. Thus, both advantages, structured naming schemes according to MTConnect and the scalable mapping of OPC UA, can be combined in terms of the companion specification.

Finally, the Automatic Identification (*de: Automatische Identifikation und Datenerfassung*) (AutoID) standard integration enables functionalities such as Radio-Frequency Identification (RFID) to function in cooperation with OPC UA. The AutoID specification provides means and standard procedures for an automatic identification and propagation of information from the field by means of bar codes, 2D codes and Real-Time Location Systems (RTLS) (Weinlaender, 2015). The integration of AutoID as a rather technical implementation than model standard shows the wide area of applications that are made possible through the scalable OPC UA metamodeling standard.

The incorporation of companion specifications shows that OPC UA offers great potentials in terms of scalability. Not only integration of ontologies or knowledge graphs that map the meaning of technical terms into a consistent information model can be realized, but it is also possible to derive whole new concepts that enable an object-oriented description of arbitrary components that are part of the system. Thus, the information modeling capabilities are also suitable for the mapping of Cyber-Physical Production Systems (CPPS) entities that contain data from the real world and propagate information within a digital context at the same time. Such CPPS or digital twin can be represented by an agent or holon that is part of manufacturing resources and is able to communicate with high-level systems at the same time.

Thus, a *companion specification for the representations of agents and MAS* would be suitable to combine the explicit modeling of shop floor components with a modeling of intelligent entities that exist in both worlds and are able to transfer information from and to these worlds seamlessly and without a loss of context information. Before the architecture and implementation of MAS enabled by an OPC UA metamodel will be presented in chapter 4 and 5 of this work the next section will cover existing approaches to wrap intelligent agents or CPS by means of OPC UA information modeling.

3.9.9 Realization of Agent-based and Decentralized Intelligence Approaches through OPC UA

As pointed out earlier, the OPC UA standard has been developed with prior purpose to industrial automation, however its applicability reaches beyond the borders of manufacturing. Thus, in order to cover existing approaches that intend to wrap the behavior of intelligent agents using OPC UA, further application domains are discussed in this section. After covering the usage of semantic web services for *Smart Grids* through OPC UA other application domains are taken into account, such as public transportation systems, inventory management before the monitoring of control systems by means of OPC UA enabled agents is presented.

Semantics for Smart Grids

An important field of action that is very active in terms of OPC UA based information for enabling intelligent systems is represented by *Smart Grids*. The term hereby describes an intelligent utilization of electrical networks. There is no unified definition for *Smart Grids* as the term can be described from different viewpoints (Rohjans et al., 2011; World Economics Forum, 2009). According to these different definitions, the term can either be interpreted as "the state of the future energy systems or as the process to realize the energy system of the future" (Rohjans et al., 2011). However, in most of these definitions Information and Communication Technologies (ICT) represent a key component of future Smart Grids. A suitable definition for this purpose might be:

A smart grid is an electricity network that can integrate in a cost
efficient manner the behavior and actions of all users connected
to it – generators, consumers and those that do both – in order
to ensure an economically efficient, sustainable power system
with low losses and high levels of quality and security of supply
and safety (definition given by the Expert Group 1 of the EU
Commission Task Force for Smart Grids). (European Committee
for Electrotechnical Standardization, 2012)

Similar to industrial production, the energy domain is facing major
changes, especially initiated through a transition from traditional supply
chains with centralized energy generation to multi-dimensional, distributed, highly dynamic and complex systems (Rohjans et al., 2011). Many
different service providers participate in intelligent energy network infrastructures that are characterized increasing information exchange, especially
between energy-specific devices and Energy Management Systems (EMS),
Distribution Management Systems (DMS) and SCADA (NIST, 2010).

Due to these requirements imposed to future-oriented electrical networks,
SOA approaches are recommended for the realization of a common concept
of multi-utility management (SMB Smart Grid Strategic Group, 2010).
Although the concept of SOA is an architectural approach that enables an
interconnection of the different systems mentioned above, it is no out of the
box solution, but rather a system that is carried out by means of individual
purposes. Thus, the goal is to provide flexible patterns that allow developers
to realize the communication between several distributed systems. Hence,
"OPC UA seems to fit perfectly into this big picture because it meets most
of the requirements. On the one hand, it is conceptually designed as SOA,
and on the other hand, it provides a suitable information architecture as
recommended standard for smart grid implementations".(Rohjans et al.,
2011)

In order to realize these concepts for intelligent systems autonomously
communicating through OPC UA a semantic web services approach was
proposed by Rohjans et al. (2010) that combines CIM approaches with the
OPC UA metamodel. The abstract architecture that has been carried out in
terms of this approach is referred to as Semantically Enabled Service-oriented
Architectures (SESA) (Rohjans et al., 2011) that attempts to incorporate
OPC UA services into the context of its rich service architecture. The "SESA
framework represents an envisioned improvement of SOA empowered by
adding semantics as a means to deal with heterogeneity and mechanization

of service usage" (Rohjans et al., 2011; Fensel et al., 2008). The result of incorporating these models with OPC UA consist in a Web Services Modeling Ontology (WSMO) for an integration of CIM semantics in terms of an OPC UA metamodel. Several OPC UA services are furthermore incorporated in a Web Service Execution Environment (WSMX), e.g. the *Discovery Service*, the *Find Servers* and *Register Service* of OPC UA (Rohjans et al., 2011). Further implementations of OPC UA with SOA approaches, e.g. in terms of DPWS have been given in Sucic et al. (2012) and Sucic et al. (2014).

Through the automatic mediation of these services, it is possible to realize autonomous method calls in an OPC UA infrastructure, e.g. by means of software agents representing the different systems on the electrical field level. A realization of OPC UA based agents by means of the semantics described above has been presented in Schütte et al. (2013). The proposed architecture enables agents to communicate through OPC UA and to use OPC UA based services to realize rescheduling activities in process. By means of the provided domain ontology, OPC UA ensures a consistent information exchange and provides up-to-date information for all participating entities.

Further Application Domains of Intelligent Agents Enabled by OPC UA

Another field, in which OPC UA has been applied in combination with MAS concepts is the Inventory Operations Management (IOM) and warehouse logistics systems (Maka et al., 2011a). In terms of an intelligent IOM Cupek et al. (2015) proposes a service-oriented communication interface model for the application of agents designed for logistic operation. The approach is of special interest as it takes into account "existing manufacturing standards and reflects the system requirements that have been defined for cyber-physical systems that are developed in accordance with the model defined by Industry 4.0" (Jazdi, 2014; Cupek et al., 2015). The behavior of the MAS is enriched by an interconnection to an MES by means of the ISA-95 standard. The according autonomous mobile platform is carried out by the introduction of several agent types, such as logistics agents, delivery agents, transportation and routing agents. Several services of OPC UA are used for the communication between these different agents in terms of embedding OPC UA server and client instances onto the physical machines of the agent entities. In this context, OPC UA information modeling capabilities are mainly used to map represented information of the agents and the platform in an object-oriented manner as well as to map transportation orders on OPC UA type definitions.

An additional field of action, in which OPC UA enabled agents are deployed, consists in public transportation systems. Inspired by the afore-mentioned inventory and logistics approaches, use-cases for an OPC UA object-oriented modeling for public transportation systems can be found in Maka et al. (2011b). In the derived domain ontology customized type definition for public transportation entities such as a *BusType*, a *LowFloor-BusType* and a *BusLocationType* have been derived in order to interconnect all physical entities influencing the information about actual states of the transportation system. All instances of the system that are derived by these type definition are represented by intelligent agents enabling a service distribution among the entities of the network.

Multi-Agent Systems and Learning Agents Enabled by OPC UA Domain Ontologies

Finally, in the domain of hierarchical control and monitoring systems, sig-nificant works have been carried out by Tan et al. (2009). The usage of OPC / OPC UA within the proposed architecture is motivated by the capab-ilities of OPC UA in terms of continuous-event and discrete-event processing through OPC UA based web services. The authors further describe that an "incorporation of XML [is used] for the negotiation and cooperation with the multi-agent system's environment" (Tan et al., 2009). In order to under-line the applicability of OPC UA in the context of monitoring and control systems, the authors define a complex control system, i.e. a process auto-mation system, as a "distributed and integrated monitoring, control, and coordination system with partially cyclic and event-based operations. Its control functions can be divided into continuous, sequential, and batch con-trol." (Tan et al., 2009) Furthermore they state that the main difference between continuous control systems and e.g. discrete manufacturing con-sists in the additional functions a complex control system has to manage, i.e. in terms of performance monitoring, condition monitoring, abnormal situation handling and reporting. Due to the distributed nature of these control systems the "main challenge is the appropriate synthesis of flex-ible network-based computer and control-systems" (Tan et al., 2009). They state that OPC UA delivers the required functionalities by providing "a paradigm for the design and implementation of control software".

Starting point for the architecture proposed by Tan et al. (2008) are the works of Dong et al. (2005) proposing an architecture for a multi-agent system for industrial process control, which consists of three layers: (i) a *cooperation layer*; (ii) a *planning layer* and (iii) a *control layer*. The control layers represents reactive behavior by means of real-time characteristics, the cooperation layer intends to provide suitable interfaces between the MAS and other systems and the planning layer contains the main functionalities such as *plan, negotiate, consistency, interfaces, information services*, etc. Similar multi-agent system approaches have been proposed by Najid et al. (2002) (multi-agent system approach with distributed hierarchical control for the flexible manufacturing cell), Damba and Watanabe (2007) (multi-agent system with a hierarchical representation designed for efficient control including individual and social criteria) and Kendall and Malkoun (1997), who presented a design for a multi-agent system using object-oriented design patterns while taking into account agent concurrency, virtual migration, collaboration, and reasoning (Tan et al., 2009). According to Tan et al. (2009) the drawbacks of the existing approaches consist in their limited ability to address monitoring operations (Seilonen et al., 2005) and accordingly to react in a real-time fashion.

Another limiting factor of the existing architectures is that the approaches are not able to establish complex interrelationships between various control variables. This lack of interrelations "makes it [generally] difficult to find problem decompositions that are suitable for agents" (Tan et al., 2009; Bussmann et al., 2001; Seilonen, 2006).

The architecture that is finally proposed by Tan et al. (2009) makes use of real-time and non-real-time applications that become possible through an extensive usage of OPC UA. The architecture consists of multiple levels agents:

- The control agents are directly connected to control objects,

- the supervisory agents directly above have the purpose of general supervision of the process performance and play an important role in cooperation, planning and reasoning based on data from the control agents,

- an OPC (Classic) based hierarchical structure is used to organize the agents by making use of a "self-organizing" database.

- Finally, an OPC UA based interface to this database enables connectivity to a broader network infrastructure and hereby enables access to export agents.

Although the proposed system architecture for MAS in complex control systems makes use of the OPC UA services and flexible communication patterns the hierarchical organization and structuring of the agents is realized by means of the OPC UA *AddressSpace*, modeling of the agents as well as their communication is still based on proprietary solutions with limited scalability. A realization of MAS that are able to make use of the full semantic stack of OPC UA including companion specifications, domain-specific ontologies and integrated communication between OPC UA based agents is presented in chapter 5 of this work after the general requirements and specifications of such architecture are derived in chapter 4.

4 Architecture of a Framework For Real-Time Interoperable Factories

Considering the challenges that are connected to the complexity of current manufacturing systems as pointed out in the problem description and fundamentals, it is the aim of this chapter to outline an architectural approach that takes into account these requirements. The challenges in terms of the derivation of an agent-based/HMS inspired architecture regarding novel manufacturing systems are very well pointed out by Shen and Norrie (1999), Jin-Hai et al. (2003), McFarlane and Bussmann (2003) and Botti and Giret (2008): enterprise integration, distributed organization, heterogeneous environments, interoperability, open and dynamic structures, cooperation, integration of humans with software and hardware, agility, scalability, and fault tolerance. The boundary conditions made up by these challenges as well as the technological approaches pointed out in the state of the art section of this thesis constitute the cornerstone for this architecture.

The arising architecture accordingly takes into consideration demands of shop floor automation and enterprise requirements at the same time. The shop floor demands are met by a MAS approach that enables autonomous decision making at the lowest level of the production in tight collaboration with state of the art automation systems. The requirements of the enterprise level are considered by enabling a scalable communication between the designed MAS and the high-level systems that are employed for production planning by means of manufacturing execution and ERP systems. The derivation of such architecture is divided into the following steps:

- In section 4.1 the requirements of current manufacturing automation systems are pointed out to determine the demanded horizontal interoperability needs of existing shop floor devices in connection with multi-agent systems on the same level.

- In the next step, a first draft of the agent architecture is presented. The architectural approach described in section 4.2 intends to outline the implementation steps needed to develop an MAS on the shop floor

© Springer Fachmedien Wiesbaden GmbH, part of Springer Nature 2019
M. Hoffmann, *Smart Agents for the Industry 4.0*,
https://doi.org/10.1007/978-3-658-27742-0_4

that seamlessly communicates with legacy automation systems such as programmable logic controllers and other low-level devices.

- Section 4.3 accordingly drafts an overall architecture for combined agent-based/HMS inspired autonomous control solutions.

- In section 4.4 the behavior of smart agents on the shop floor to enable autonomous decision making and flexible production organization will be described based on the derived architecture.

- The last section of this chapter will focus on the limitations of the derived approach. The sole architecture enabling agents on the shop floor may not be sufficient to reach the desired interoperability goals. Thus, section 4.5 describes the drawbacks of using legacy communication approaches to realize interactions between software agents and traditional automation systems and higher systems of the factory hierarchy.

The following chapter describes the derivation of concepts that enable a natural communication between software agents of the proposed approach.

4.1 Requirement Analysis for Legacy Systems in Automated Production Sites

The requirement analysis for integrating legacy systems into modern and flexible manufacturing approaches directs to several different perspectives. Current production organization is primarily determined by means of the management perspective (section 4.1.1) focusing on the decision-making with influence on the entire corporation. On the levels below the strategical decision layers, engineering requirements in terms of upgrading legacy automation systems are in focus (section 4.1.2). The requirement analysis is closed by taking into consideration the information technological demands of an architecture that fits the needs of a flexible automation system (section 4.1.3). These requirements open up perspectives on how to bring together management requirements and engineering demands from an information technological point of view, but also point out the limitations of considering those different paradigms at the same time.

4.1.1 Management Perspective of Automation Systems

When we look at current automation systems, the implementation of process operations is based on fixed schedules and production programs. This kind of process organization is due to the management-driven approaches of current factories that require a production process in the most efficient way. As already mentioned in section 3.2.1 the managerial goal is mostly related to the reduction of waste and in terms of increasing the efficiency of all processes.

From this top-level point of view, production management "deals with decision making related to production processes so that the resulting goods or services are produced according to specifications, in the amount and by the schedule demanded and out of minimum cost" (Buffa, 1983; van Biljon, 2004; Metz, 2014). The goals of this manufacturing requirements is to produce goods according to the right quality and quantity, in the right time and with appropriate manufacturing cost (Kumar and Singhal, 2014; Metz, 2014).

The information that is collected during the production process accordingly has the single goal of serving requirements based on KPI and similar performance figures. The decisions that are made on the management level concerning production goals are strongly influenced by the decision relevance, by the amount of aggregated information available for decision-making and the prospective horizon of the decisions to be made (Günther and Tempelmeier, 2012). When further examining the production organization in an enterprise based on the managerial perspective, three aggregation layers have to be distinguished – the strategical, tactical and operational layer targeting different goals and representing different information needs to fulfill their aims.

Thus, the first aim of the management perspective is to fulfill the strategical goals of the enterprise rather than concentrating on the lower levels that concern the tactical and operational planning of the production. Not until these high-level strategical decisions were made, the planning concentrates on the reorganization and further development of the infrastructure and production organization (Günther and Tempelmeier, 2012).

The tactical planning and organization of the production is directly connected to the control of cost and profit (Grauer et al., 2011). This includes cost accounting, performance evaluation of the process in terms of KPI, cost-utility analysis and the provisions of means to estimate the cost effectiveness of production factors (Piontek, 2005). Summarizing the aims of the tactical decision layer of the enterprise, it targets the past and future manufacturing activities in terms of costs and performance.

The operational planning layer of the enterprise is characterized by decision with regard to the current situation of the production. The *supervisory management* undertakes operational decisions that are located at the interface of the actual manufacturing process and the allocation of resources (Günther and Tempelmeier, 2012; Metz, 2014). All operational decisions that are focused on the Production Planning and Control (PPC) are taken this operational level (Fandel et al., 2011). In most cases, PPC is based on enterprise resource planning concepts represented by ERP systems and in cooperation with Material Requirements Planning (MRP) and capacity planning as well as (re-)scheduling (Groover, 2013; Pinedo, 2014).

Besides ERP systems, there are also other attempts to perform operational planning from the management levels, e.g. by means of Manufacturing Resource Planning (MRP II) concepts. Introduced by Higgins et al. (1996), MRP II are capable of representing different levels of the production planning, i.e. in terms of strategic, tactical and operational planning. In accordance to the proposed concepts, detailed (re-)scheduling, dispatching and shop floor control can be interpreted as operational decisions based on the MRP II concepts (Metz, 2014). However, other works argue that using the proposed methods of organizing the operational planning from the management level it is not sufficiently possible to cover the most important requirements using current ERP or MRP II systems (Kurbel, 2016).

These shortcomings of high-level systems when it comes to managing resource and operational planning on the shop floor have led to the introduction of MES as an intermediate layer between top-level and field-level systems, as MES are in direct contact to manufacturing facilities like machines or resources and provide suitable interfaces to ERP systems at the same time. The utilization of MES covers the requirements of our desired architecture from the engineering perspective.

4.1.2 Engineering Perspective for the Requirements of Flexible Automation Systems

Due to the lack of accurate up-to-date information from the field level as the basis for fundamental managerial decisions, most of the decisions made on the management level accordingly focus on mid-term and long-term goals as these can be appropriately evaluated using the level of available information. Short-term goals like prioritizing certain products or an optimization of single process goals are not in focus of these decisions, because they are not of strategical relevance for the enterprise and more important – current information bases are not able to provide an appropriate knowledge base for low-level related decisions.

The lack of data from the field as described above characterizes the main drawback of managerial decisions for manufacturing purposes, which is primarily based on the missing feedback loop between process planning and the actual manufacturing loop (Metz, 2014). Data in the field is primarily recorded for the diagnosis of errors or production faults. However, the granularity and nature of this data does not fit the needs for accurate decisions in most cases as the information exchanged lacks context and other descriptive information for the data that is collected.

Thus, in order to successfully take into account the engineering perspective in terms of planning and reconfiguration of the manufacturing process, the MES concepts can be utilized to bridge the gap between the information systems of high-level enterprise application and the information technological needs of tightly coupled systems on the shop floor. The integration of MES is primarily focused on enabling control loops between the ongoing processes in the field and the according control instances, i.e. programmable logic controllers, in order to reach a higher overall process efficiency characterized by quality, timeliness and maintenance aspects. A typical structure of such organizational approach is depicted in Figure 4.1.

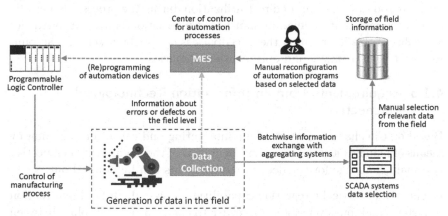

Figure 4.1: Production organization in current automation systems

On the shop floor a robot is shown as representation of a field level device. The automated processes of this robot are controlled by the programmable logic controller that contains all necessary programs, sequences and logical correlations for the manufacturing steps performed by the robot. This fixed manufacturing program of the robot can be disturbed for example by material faults or other external influences. When such errors occur, some diagnostics data in the form of fault bits or similar rudimentary signals will be processed to the manufacturing execution systems signalizing that there might be problems during the production process. Based on these signals the manufacturing execution systems may stop the process or reinitialize the production based on human interaction with the system configuration.

Besides this automatic data exchange between the field and managing instances of the automation system, there is also some manually performed information aggregation of certain field level data. This information that might be collected for diagnostics and analytical reasons during the manufacturing process is shown within supervisory control and data acquisition systems. The information that is available in such SCADA applications usually contains aggregated information about the ongoing process, such as average time for production steps, deviations in terms of manufacturing intolerances or similar KPI. This information that can be propagated into higher systems or into databases serves as the basis for later decision-making in terms of tactical or strategic decisions concerning the production planning and control. However, the cycle time of these feedback loops that contain information about the field with major relevance for high-level systems are too slow in terms of direct utilization during the process. Especially due to the fact that manual, human-based interaction is needed to extract the value of this information these feedback is cannot be taken into account for short-time planning and optimization scenarios.

4.1.3 Requirements from an Information Technological Perspective

Based on the characterization of manufacturing and execution planning by means of management decisions and by means of the engineering perspective two main challenge for the desired requirements can be pointed out:

- The lack of vertical integration capabilities between high-level information systems and low-level systems in order to enable a tight coupling between strategic decisions and the actions that are effectively performed on the shop floor.

- The realization of (near) real-time multiple closed control loops that accurately describe the demands of an ongoing process on the shop floor (Metz, 2014).

In order to meet this challenges with suitable concepts the system components of the manufacturing hierarchy need to be evaluated appropriately in terms of their information technological specifications. Figure 4.2 summarizes the requirements from the information technological perspective in terms of the discussed systems. As shown in the figure, an information exchange between the low-level systems and the operational, tactical and strategical levels could contribute significantly to vertical integration and to realize control-loops between the shop floor and the systems of PPC.

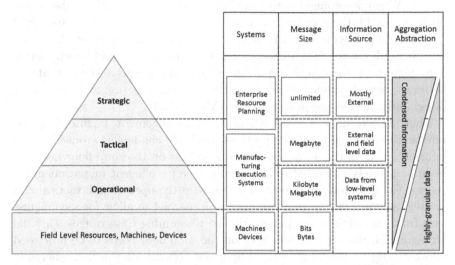

Figure 4.2: Requirements of modern interconnected automation systems from the information technological perspective, according to Metz (2014)

The levels of the pyramid are characterized by the presence of different systems. The ERP systems usually constitute the interface between the strategic and tactical levels of the enterprise, whereas the MES are more located on the edge of the tactical and operational levels. According to these system configurations the vertical hierarchy of the pyramid can be covered by systems according to a top-down topology from the strategic enterprise planning to the shop floor containing the process flows of the manufacturing.

The systems that are present on the different levels of the hierarchy are furthermore characterized by various requirements in terms of message size, information source and abstraction level of the available information. Due to the different granularity of information that is manifested regarding message size, the data on the vertical levels is of high heterogeneity.

The messages or data transactions that are performed on the shop floor, are usually of high frequency, low size and characterized by various context information. This leads to a high granularity of data, massive amounts of data sets and loads of detailed information about the ongoing processes. In order to process this data, i.e. interpret the information that is connected to this data sets, these high amounts of data need be condensed to some extent. When accordingly aggregating the information into the higher levels of the pyramid, the context information that was connected to the initial data sets usually gets lost, which makes it difficult to interpret the initial meaning of low-level data sets on higher levels of the pyramid appropriately. This issue gets even more significant when taking into account that the heterogeneity of systems increases in the vertical direction and gets closer integrated near the physical resources (Metz, 2014).

Thus, as a result of the requirements of management, engineering and information technological perspectives, multiple challenges arouse: (i) the information from the actual production process on the shop floor needs to be appropriately processed to be interpretable for efficient management decisions; (ii) a vertical interoperability between the operational, tactical and strategical levels of the company has to be reached to allow for reconfiguration from the engineering perspective in reasonable time ranges; (iii) the high-granular data that is generated on the shop floor needs to be processed appropriately in order not to lose important context information inherently present within the data from low-level devices. Accordingly, an architectural approach that intends to represent a next level solution for highly integrated automation systems needs to address these requirements in an appropriate way.

4.2 MAS Architecture Enabling Interoperability with Existing Automation Solutions

The architectural approach that is presented in this section, attempts to cover the requirements and possible issues that were aroused in the previous section. This endeavor is performed by introducing a system architecture that works closely integrated with the shop floor processes, but at the same time enables suitable interfaces to provide aggregated information to the decision-making layers of the production hierarchy. The proposed architecture is based on the MAS approaches covered in section 3.6 and additionally covers elements of HMS.

4.2.1 Basic Architecture

The basic architecture is oriented towards the system capabilities of manufacturing control systems that are composed of software modules as well as different physical elements of the manufacturing environment (Botti and Giret, 2008). Besides resources and products that are present in every manufacturing system, production control systems are also characterized by coordination operations, resource allocation or work orders from clients. In order to deal with these different nature of required resources and their according representation on the shop floor, various tasks need to be allocated onto different entities that represent various physical and/or software components. Physical components are usually represented by manufacturing resources as machines, devices for the gathering of data like sensor or other kind of actuators. The software components embedded into the field need to be capable of decision making as well as communication capabilities.

In terms of the paradigms in holonic manufacturing systems, "the software module and the physical entity, bonded by means of an appropriate communication network, represent a holon in [the] manufacturing system. Each of these holons will be able to reason, make decisions, and communicate interactively with other holons" (Botti and Giret, 2008). The systematics of interacting holons, the number of holons with different capabilities as well as their configuration define the basic architecture of the HMS. One of the first HMS was presented by Christensen (1994) in terms of the architecture shown in Figure 4.3.

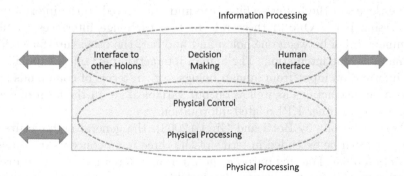

Figure 4.3: General architecture of a holon

This architectural approach covers both parts of the aforementioned holon components, which are represented by a part that is capable of processing physical behavior of the automated manufacturing systems and other part that is responsible for information processing and communication. The physical processing part of the architecture is an optional element as it only constitutes an extension of the vital parts of decision making and communication. Examples of holons that do not rely on a physical component are work-order holons, planning holons or scheduler holons (Botti and Giret, 2008). As these kind of holons are commissioned only for organizational tasks, no physical interaction nor physical processing is regarding their responsibilities. The physical component of the holon is further divided into two parts: The physical processing itself representing the hardware that executes the actual manufacturing operation; the second part consists of a physical control unit, i.e. a controller by means of a PLC or CNC unit. The information processing component of the holon is comprised of three parts: The central decision making component that contains the capabilities for reasoning (Botti and Giret, 2008) as well as interface related components.

The general architecture also shows interaction capabilities with the surrounding environment: (i) On the lower level, the physical processing part is in direct contact with the manufacturing process accordingly exchanging measurement variables or similar process related figures. This interface actually represents the concept of CPS quiet well as the physical processes are transformed into the digital world by means of an entity that coexists on the physical layer as well as on the edge of the digital representation at the same time. (ii) The other interaction capabilities are located within the higher levels of the holon architecture and are related to the information processing part. One of these information exchange interfaces enables communication between one holon and another, the other interface offers human-machine interaction. The human interaction part offers capabilities for information input from human beings, e.g. in terms of commands and operation instructions as well as output variables, i.e. in the form of state monitoring or other KPI related information.

As pointed out by Botti and Giret (2008), the general architecture of the holon can be extended in terms of the three main concepts mentioned in section 3.6.5. For this purpose the internal holon structure in terms of decision-making needs to appropriately address these main requirements in terms of autonomous control of machinery, cooperation with other holons and interaction within multiple organizations. The following infrastructural extension takes into account these main objectives by further specifying the internal holon structure in terms of decision-making and inter-holon communication (see Figure 4.4).

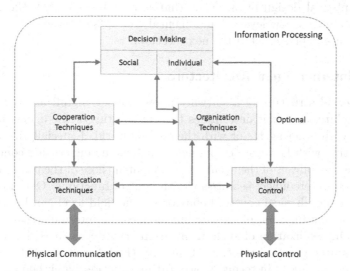

Figure 4.4: Internal holon architecture for information processing

The decision-making module is separated into two parts, which are related to both social and individual techniques. The individual decision-making in terms of personal goals and strategies is strongly influenced by the current situation from the physical world. Accordingly, the decision-making module of the holon makes use of the Behavior *Control module* that is in charge of the physical control interface. Through this interface the decision-making module is able to stay informed about the current situation and to translate behavior control actions of the holon into hardware operations (Botti and Giret, 2008).

In terms of social interaction, the decision-making module makes use of cooperation techniques with other holons. These cooperation techniques access the communication techniques, which represent the knowledge about different domain ontologies and languages. Other interaction capabilities in terms of cooperation are represented by *organization techniques* that contain means for reorganizing the manufacturing using process control. By accessing the *behavior control* module the holons are able to figure out opportunities for improvement and can start negotiation processes with other holons in order to initiate the reorganization of the manufacturing (Botti and Giret, 2008).

After these clarifications in terms of the information processing behavior of holons respectively agents the next steps to a suitable architecture consists in the internal design of an agent that is able to deal with the process information from low levels. Fundamental concepts of these internal agent architecture is presented in the next section.

4.2.2 Internal Agent Architecture

The internal structure of a software agent plays a significant role in the MAS architecture at it determines the capabilities of the agent interconnecting with other agents or with the outside world. Especially in terms of cooperating with hardware components such as manufacturing control systems the capabilities of the agents need to match those of the physical world in order to interconnect seamlessly with control functions. One possibility to match the physical control behavior on the field level can be achieved by integrating function block capabilities into the internal structure of an agent. One realization of such agent architecture was carried out be the HMS research group of the Keele Univerity (Fletcher et al., 2000; Fletcher and Brennan, 2001) in terms incorporating function block behaviors into the MAS. A formal structure of such internal agent architecture is depicted in Figure 4.5. The idea behind the architectural concept of the research group is to add a hardware component in addition to the traditional agent components that comprise of knowledge and software modules (Botti and Giret, 2008).

The structure of this internal agent architecture consists of a *head* that contains high-level controlling systems of the agent. In addition to these cognitive agent functions, the *base* part represents the agent's processing systems for low-level functions, e.g. as they are used in low-level controllers like PLC. The *head* structure is inspired by the basicHMS architecture of Christensen (1994) and is made up of the according modules. These modules represent all functions that are necessary for interacting with low-level systems of the manufacturing as well as with high-level planning systems and HMI (Botti and Giret, 2008):

PMC The *PMC* module represents the process/machine control and accordingly executes controlling plans of the running processes.

PMI The *PMI* module constitutes the process/machine interface and provides the logical and physical communication interface for the processing system.

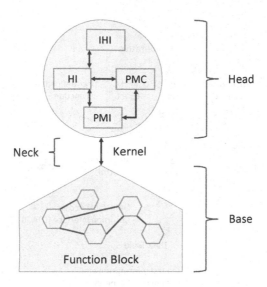

Figure 4.5: Internal holon architecture with low-level function blocks

HI The *HI* represents the human interface and provides information in a human-readable form and accordingly allow for intervention through decision-makers.

IHI Finally, the *IHI* module consists of the interholon interfaces and enables the communication and functional interaction between holons.

The function blocks in the low-level part of the architecture represent the control functions that are standardized by the IEC. The standards for the function blocks designed for manufacturing automation comprise of the architectural approaches and required software modules and can be found in IEC 61499 (2000, 2001). The main difference between the functions of the head structure and the base of the holon model consists temp range reference. While the high-level functions work in terms of loosely coupled systems that are interconnected through services, the functions of the base structure aim at managing real-time process/machine-low level control (Botti and Giret, 2008), thus representing tight-coupling concepts of the shop floor.

An advancement of this architectural approach for a single agent that fits the requirements of extended software modules and physical interaction well was developed by Brennan et al. (2003) and Brennan and Norrie (2003) with the holonic architecture for device control (HDC) (see Figure 4.6).

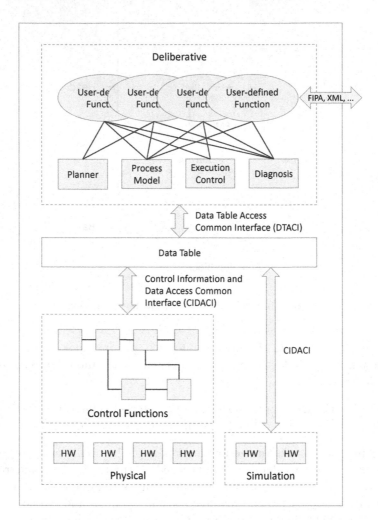

Figure 4.6: The HCD architecture according to Brennan and Norrie (2003), figure after (Botti and Giret, 2008)

The architecture is composed of two parts, the deliberative layer for organizational and administrative functions as well as function blocks for the control of physical hardware and devices. The deliberative layer on the top of the agent architecture is in charge of multiple user defined functions or control mechanisms such as planning, process modeling, execution control or diagnosis of the manufacturing. These functionalities are subsumed as *generic functionalities.* Further functions of the deliberative layer may be in charge of application-domain-specific functionalities in accordance to the user-defined functions. The control functions being part of the function blocks layer are in charge of hardware control and usually consist of user-defined functions and control loop definitions. The according actions as well as automation processes are realized by means of function blocks. Below these function blocks the physical layer represents direct interaction with the hardware, whereas the simulation blocks are used for learning purposes in terms of simulating hardware components.

The internal communication of the deliberative layer with other parts of the holon is realized by means of the device data table through the *Data Table Access Common Interface (DTACI)*. Both, the deliberative and the control layer can read and write from and to the DTACI (Botti and Giret, 2008). This way, the high-level functions in terms of the deliberative layer as well as the low-level control function for hardware devices are able to work together according to a fixed set of information exchange mechanisms.

The communication capabilities of the deliberative layer with other agents is characterized by data exchange mechanisms based on proprietary messages in the form of XML syntax and according to MAS standard communication models for example defined by the FIPA. The utilization of communication standards like the one designed by FIPA is a first step towards standardizing the application scenarios of holons in manufacturing systems.

An architectural approach that again focuses more on the different knowledge dimensions a functional MAS needs to be capable of had been carried out by Fischer (1998). The basis for this approach is the three concurrent layer agent architecture INTERRAP of Muller (see Figure 4.7). The straight arrows within the figure represent control flows while the dotted arrows indicated information exchange. Each functional layer of the architecture is characterized by a distinctive sort of knowledge that focuses on the needed capabilities in terms of the different tasks of each layer.

The INTERRAP architecture had been originally developed to meet the requirements of modeling dynamic agent societies, e.g. for robot environments. The main feature of the architectural approach consists in a combination of behavioral patterns with explicit planning facilities (Müller and Pischel, 1993). Thereby, the pattern of behavior are implemented to enable the agents to quickly react on environmental changes in a flexible manner, whereas the planning pattern are focused on solving highly sophisticated problems. The architecture builds upon works that attempted to structure a knowledge base that reflects the complexity of the knowledge

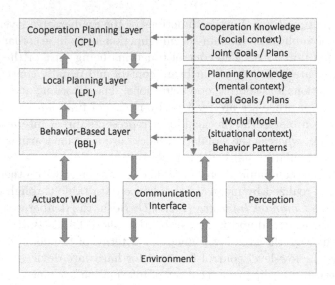

Figure 4.7: The INTERRAP architecture acc. to Müller and Pischel (1993)

contained. The main critics point in these early attempts, however, was that most system architecture for knowledge bases did not properly consider a separation between *aspects of knowledge*. They layered architecture that is proposed by the INTERRAP model makes a clear distinction between the pure knowledge base and the functional parts and at the same time preserving the hierarchical structure of earlier models.

4.2.3 HMS Reference Architecture

After characterizing the basic internal and external architecture of an agent in the previous section, the reference architecture of HMS is explained in the following. Based on the functional behavior of agents and the reference model representing a composition of multiple agents into holonic/agent-based systems the architectural requirements for the architecture of this work can be derived.

The HMS reference architecture from van Brussel et al. (1998) provides essential parts for the basic infrastructure of a functional manufacturing system that relies on the agent, i.e. holonic concept. The structure of the HMS reference architecture is constructed by means of three basic holon types: order holons, product holons and resource holons (van Brussel et al., 1998). Figure 4.8 depicts the essential structure of the base holon types.

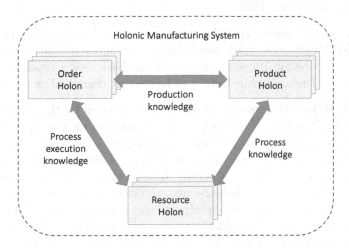

Figure 4.8: Basic holons types of the HMS reference architecture (van Brussel et al., 1998)

Instances or specializations of these holon types are capable of forming a functional manufacturing environment with all needed process and organizational related capabilities. The specification of these basic holon types as well as their further development for high-level planning capabilities can be performed in an object-oriented way, i.e. the basic holon types serve as an extensible foundation for the evolution of flexible manufacturing and automation systems.

The reference architecture of van Brussel et al. (1998) also proposes the expansion of the agent system by so-called staff holons that are capable of expert knowledge. The three basic holons present in the reference architecture can be further characterized by the fundamental requirements that are connected to the management and automation of a manufacturing system (Bongaerts et al., 1996):

1. The management of resources and according aspects such as programming and driving of machines at an optimized level of speed to maximize manufacturing capacities,

2. production related technological aspects with regard a selection of optimal parameters for achieving products of high quality,

3. and finally aspects related to the logistical organization of a manufacturing process with close interaction to customer demands and due dates (van Brussel et al., 1998).

Based on these holon types the manufacturing and automation process is organized. Unlike most traditional shop floor control architectures, as for example described in Bauer et al. (1991), the basic holon types, which are part of the reference architecture, can be further described as follows:

Resource Holon The resource holon is characterized by the physical part of the manufacturing component, e.g. a machine or some other resource entity that performs production-related actions. Resource holons are further capable of representing abstractions of the production, such as a factory, machines, furnaces, conveyors, pipelines, pallets, components, raw materials, tools, tool holders, material storage, personnel, energy, etc. Besides this production resource part, the resource holon also contains an information processing part that controls the actions of the according hardware entity. Regarding the relations to other holons, the main tasks of the resource holon are the management of production capacities and also to offer capabilities of their resource to the surrounding holons (Wyns et al., 1996). The resource holon further contains the mandatory methods for allocating resources to the according production steps as well as background knowledge about how to "organise, use and control these production resources to drive production" (van Brussel et al., 1998).

Product Holon The product holon is in charge of the production process besides the manufacturing itself, i.e. containing knowledge about the correct making of the product in sufficient quality (van Brussel et al., 1998). This knowledge is accompanied with up-to-date information about the product life cycle and other *product model* related information, such as user requirements, design, process plans, bill of materials or quality assurance procedures. Accordingly, the product holon condenses the knowledge that is traditionally covered by company divisions which focus on product design, process planning and quality assurance (van Brussel et al., 1998). The product holon acts as a sort of information server or information source to other holons that part of the HMS and require detailed knowledge about organizational or production related information.

Order Holon The product holon matches a product in the form of an order, thus represents the flow of one customer order or similar transaction through the factory. The order holon manages the flow of a product through the manufacturing by taking into account the product state model, as well as the processing of logistical information related to the production job. Besides customer order, the product holon may also represent make-to-stock orders, prototype orders, orders related to maintenance or repair services among others. The order holon can be seen as the workpiece managing the production of itself in terms of certain control behavior that is related to the product specific goals. In this way, the order holon is able to negotiate in the name of the product. The order holon performs tasks that are traditionally assigned to systems such as a dispatcher, a progress monitor or a short term scheduler (van Brussel et al., 1998).

As shown in Figure 4.8 the different holon types add up to each other by sharing certain kind of knowledge. The *process knowledge* that is exchanged between product holons and resource holons contains information on how to perform a certain manufacturing step on a certain resource. This knowledge takes into account the capabilities of the according resources as well as the requirements related to the process outcome regarding process parameters, process quality, etc. *Production knowledge* is exchanged between product holons and resource holons and combines knowledge about manufacturing processes and the product, thus, how to make use of certain manufacturing resource to make a certain product. This knowledge comprises process plans, sequences and other production order related information. Finally, *process execution knowledge* is exchanged between resource holons and process holons and contains information about the occupation of resources with manufacturing steps and the consequences of assigning production steps, interrupting them or suspending and resuming the production on a certain resource. The sharing and utilization of these different types of knowledge between single instances of holons unleashes the full capabilities of the HMS. The combined knowledge enables a fully functional manufacturing system that plans, manages and produces autonomously taking into account dynamic behavior of the manufacturing.

4.2.4 Cross-Domain Architecture

In addition to the different functional layers that characterize the behavioral patterns of an agent, the presence of multiple application domains in specific use-case has to be considered equally. This means agents may not only focus on a specific (expert) domain, i.e. the usage of other knowledge domains for pursuing their goals may be inevitable for targeting their design objectives in some cases. In order to take account for the necessity of these requirements, the concept of *cooperation domains* has been carried out by Fletcher et al. (2000).

In consideration of the theoretical background on cross-domain HMS a cooperation domain is considered as a logical space that enables holons to communicate and to operate with each other within a certain context. In other words – cooperation domains provide a cooperation metaspace, in which agents can communicate by making use of common semantics. Within cooperation domains it might also be possible that agents share common goals and target objectives.

Cooperation domains are defined in a flexible manner, thus supporting a loose coupling of the agents involved. This means that cooperation domains do not necessarily exist by themselves, but can be dynamically generated through the process and certain operations. This also implies that cooperation domains do not need to be created or named in advance, although some cooperation domains may be created to cover certain applications of the HMS. Fletcher et al. (2000) propose several premises for defining the validity of cooperation domains: (i) A holonic system contains at least one cooperation domain. (ii) A holon is a member of one or more cooperation domains. (iii) A cooperation domain has one or more member holons. Furthermore, a cooperation domain consists of the following elements:

- Suitable data structures that are used for the consolidation and retrieval of knowledge by the agents. The stored knowledge may be used for controlling the cooperation, e.g. in terms of querying a variable that indicates the status of a joint task (Fletcher et al., 2000).

- Another required element is represented by facilities to pass transient messages between holons and the cooperation domain. If the cooperation domain and the agents being part of it are located on the same physical machine the message exchange can be realized by means of a shared memory. Otherwise, the messages have to be encoded and sent through a communication stack.

- Cooperation domains need to be capable of decision making mechanisms in order to support agents in performing their tasks, especially in terms of planning, negotiation, information exchange and similar activities.

- The management of application domains needs to be capable of techniques and rules to decompose and allocate tasks among holons that share a common domain. Furthermore, the application domain needs to support the holon in scheduling and controlling its tasks.

- If tasks are distributed among several agents, the cooperation domain has to provide means to monitor the status and to schedule/control all actions within these tasks.

The generation of cooperative tasks is initiated by constituent holons that – among other holons – are part of a cooperation domain. By creating and executing these plans holons take into account their respective views of the holarchy, i.e. the hierarchical or heterarchical composition of the HMS. Every holon actively taking part in the cooperation domain planning is mapped onto one or more holonic resources in the system, where at least one of these resources has to provide management services for the cooperation domain (Fletcher et al., 2000). When cooperative tasks have been determined, information processing elements within one cooperation domain interact with each other in order to accomplish a specific task.

In order to increase the flexibility of cooperating compounds of holons, the holon term is further extended by the cooperation domain concepts. In this context, a holon may be composed of a set of other holons. The configuration of these holons might be in the manner of a recursive containment hierarchy/heterarchy – or more generally spoken: a holarchy – to form a compound in terms of a parent holon. With regard to these parent holons, their according sub holons take part in the goals and design objectives of their parents, accordingly cooperating to fulfill the tasks defined by their super types. If a holon does not contain further children, it is defined as an atomic holon. In this case, the internal cooperation domain of the holon is equal to the holon's private autonomous functions and information (Fletcher et al., 2000). This rather complex composition of cooperation domains is summarized in Figure 4.9.

The cooperation domains form spontaneously and can be composed of any number of holons. The holons within a cooperation domain are able to develop stable intermediate forms during their evolution and are self-reliant regarding their relations to other holons (Bussmann, 1998). As shown in the figure, the heterarchies can be of arbitrary depth and can be composed of a flexible number of holons. According to the concepts of these cooperation domains, every holon always shows all of its capabilities, accordingly allowing other holons to browse for their abilities. This enables complex interaction capabilities and allows for the resolution of complicated problems that require a cooperative execution of joint plans.

While the structure that is created by tasks holarchies is dynamic, the relationship among holons form a rather static configuration (Fletcher et al.,

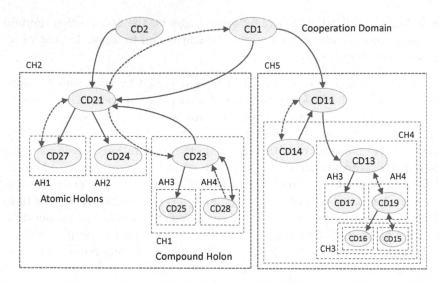

Figure 4.9: Holarchy of holons embedded in several cooperation domains (redrawn according to Fletcher et al. (2000))

2000). After task requests have been communicated into the system, holons respond from their cooperation domain in a way that new (joint) tasks can be carried out according to these responses. Due to this way of communicating possible lack of skills within a cooperation domain can be compensated by introducing new holons into the cooperation domain, if these holons satisfy the domain's requirements.

In addition to the general and internal architecture of agents, this previous section forms the last building block for the derivation of an overall architecture that fits the requirements of cooperating, autonomous agents embedded into real-world manufacturing system.

4.3 Overall MAS Architecture for Cooperating Agents in a Smart Factory Environment

Based on the internal and external architecture approaches described in the previous parts, a general concept for the overall architecture of cooperating agents in consideration of current demands in manufacturing system is carried out in this section. The architecture takes into account the internal structure of a holon – or agent – and specifies external as well as internal communication and cooperation capabilities based on the demands and requirements of automation systems. The management of hardware components and control functions for an adaptive process behavior is managed by different interfaces and interaction capabilities.

4.3.1 Multi-Agent System Architecture

The basic mapping of functions that are required for a self-sufficient MAS that is capable of organizing and executing the production of highly customized products is inspired by the reference architecture approaches, which define the required agent or holon types. These holon types, namely in form of the order holon, product holon and resource holon represent the according sources of knowledge that is required to manage the process from high-level functions to low-level execution plans.

In terms of the overall architecture that is carried out in this work, the required functions go beyond the perspective of single holons are realized in terms of a recombination of holon types and MAS concepts. This general approach provides a mapping of the typical holon characteristics to a MAS approach as shown in Figure 4.10. The nomenclature of the according agents is inspired through cooperative research activities of the author within an agent expert group on agent-based approaches and their according demonstrators, e.g. presented in works from Pantförder et al. (2014) and Vogel-Heuser et al. (2015).

The functionalities of the order holon prototype are extended in terms of a *customer agent* that is responsible for all relations regarding orders, order management and processing of customer demands. This agent accordingly represents an interface function to high-level systems such as ERP or PLM. This specialized agent additionally takes care about the propagation of customer-related information into the multi-agent system and to pass on internal customer and administrative planning related information from the MAS back to the *planning layer* and its according systems.

The *coordination agent* represents and extends the abilities of the product holon, which summarizes the functionalities of resource allocation and process control. The main (extended) functionalities of the coordination agent are the acceptance and processing of orders from the customer agent respectively the planning layer and the transformation of these abstract orders into executable manufacturing plans. Thus, the scope of action regarding the coordination agent is congruent with some main functionalities of MES. The coordination and concrete production planning performed

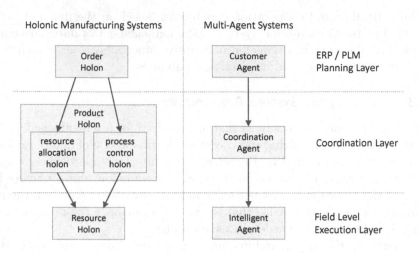

Figure 4.10: Mapping of holon type functions to MAS instances

by this agent can occur with regard to a entire production system or just regarding a certain cooperation domain. The coordination thereby assigns possible works tasks according to the capabilities of the agents within the field without suppressing their autonomy in terms of whether the intelligent holons are going to perform the assigned task or not. The relationships between the coordination layer and the executing layers are characterized by loose coupling and flexible communication hierarchies.

The *intelligent agent* that is also referred to as the autonomous agent of the MAS represents the resource holon and is located on the field level. This agent accordingly implements all functional competencies and capabilities that are required to perform the manufacturing process on the shop floor, specifically supporting the agent concepts of autonomy, cooperation and self-organization. In contrast to these agents that perform the actual work tasks regarding the desired products, the coordination agent is rather responsible for the mediation of the relations and information flows between the intelligent agents than for concrete execution.

In terms of the architecture carried out within this work, the general requirement and desired capabilities are matched precisely by the concepts of MAS as multi-agent based approaches subsume and extend the HMS concepts in several distinctive points. Although, the general understanding of MAS and HMS have many concepts in common such as autonomy, reactivity and goal-directed reasoning in order to realize individual behavior

(O'Hare and Jennings, 1996; Huhns, 2000; Müller, 1996). Despite these congruent behavioral patterns, MAS are characterized by additional aspects in terms of cooperation, coordination and negotiation with regard to coalition formation, role assignment and self-organization in order to form social behavior (Bussmann, 1998; O'Hare and Jennings, 1996). Especially the last point of these behavioral patterns allows for scalability and evolving of an agent and is therefore of high importance for MAS in flexible, adaptive environments.

Nowadays, the borders between application of HMS and MAS become more and more blurry. Whereas the holonic concepts focus rather on concrete manufacturing problems, MAS concentrate more decisively on means to solve problems in a cooperative manner. Thus, in terms of developing practical applications by means of agent-based approaches, the holonic vision can serve as a good starting point to carry out the system architecture. The extension of the arising framework can be furthermore supplied with reasoning techniques by the concepts of MAS in order to implement the required information processing architecture for holons. Finally, the cooperation techniques of MAS are able to support an interaction between holons and to build up functional holarchies. In this way, modern manufacturing systems make use of a recombination of both approaches in order to pick up the concepts required for intelligent manufacturing optimization that works according to the low-level requirements of the process and is well integrated in terms of an overall manufacturing organization at the same time. Following this approach, the architectural concepts in this work is also carried out in this way considering both the ideas and basic concepts of HMS as well as fundamental concepts of planning and negotiation through MAS (see Figure 4.11).

Each agent in the architectural outline is represented by a holon instance including its according basic functionalities that are inspired by the PROSA reference architecture that was described in section 4.2.3 (van Brussel et al., 1998). The agent paradigm from MAS inspired approaches serves as a sort of wrapper that summarizes the holon instances and its capabilities while extending its communication and coordination abilities in the fashion of MAS. Hence, this architectural approach fulfills various requirements imposed in section 4.1 and provides several advantages according to the aimed interoperability framework:

- Each agent consists of distinctive functionalities that enable for autonomous behavior within decentralized environments.

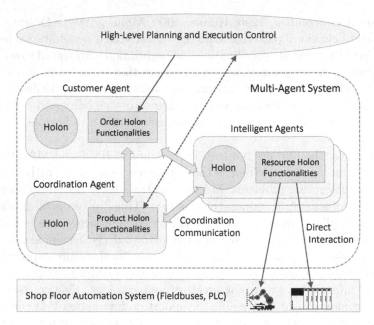

Figure 4.11: Recombination of holon and MAS concepts

- Specialized agents with high-level functionalities such as agents incorporating order holon or product holon capabilities are able to communicate with high-level information systems such as ERP, MES or PLM.

- A flexible number of intelligent agents immediately represent resources on the shop floor, accordingly aiming at direct information exchange with low-level automation systems.

- The *agent wrapper* allows for an extension of these functional requirements in terms of coordination, cooperation and communication capabilities.

- The multi-agent system shell containing all agents of the cooperative network serves as a flexible envelope summarizing all intelligent entities present within the smart manufacturing system.

The architectural approach depicted in Figure 4.11 represents an ideal target vision of an intelligent MAS in a modern production system. However, this external view on the single agents forming a functional MAS only shows general communication and cooperation flows. The internal agent behavior of the intelligent agents shown in the MAS concept have to be further detailed in order to reach the close/tight interaction of these agent entities with the hardware components as outlined in section 4.1.2. The next section will focus on the internal behavior of the agents.

4.3.2 Internal Agent Behavior and Hardware Component Interaction

This section focuses on an attempt to systematically embed an intelligent agent into the control of manufacturing units. In this context, each agent of the MAS is designed to encapsulate a distinctive unit within the production line or a process cell. A blueprint of such a *production cell* is depicted in Figure 4.12.

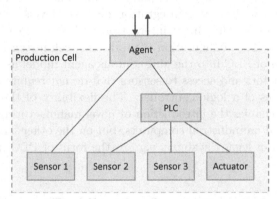

Figure 4.12: The low-level interaction blueprint of the *Production Cell*

The model shape of this approach locates the intelligent agent *on the edge* of the production cell. This placement implies that the agent is part of both worlds – high-level systems as well as other agents in terms of a loosely coupled infrastructure and at the same time manifested within the low-level hardware configuration aiming at tight coupling with embedded devices like PLC, sensors and actuators.

This approach attempts to overcome the fact that MAS solutions are nowadays for the manufacturing control despite agent-based approaches could increase the flexibility in comparison to traditional PLC especially regarding planning, executing and supervising the production process. This lack of integration in terms of MAS solutions into manufacturing systems is due to the high risk in introducing such control models, as these solutions use software which is incompatible to conventional control software – especially in terms of closed control loops and determined real-time programs on PLC devices. A closer integration into the process or with existing fixed control loop solutions has not been realized, yet, as there are no fixed/standardized communication models and interaction capabilities to these field bus controlled systems available.

If these communication and interaction capabilities, however, could be realized in both directions – to the higher levels as well as into shop floor operations – the concept of the production cell then allows for a tight interaction of the agent with the executing hardware. The agents of the architecture move more towards the shop floor without loosing their rich services in terms of high-level planning and control systems. These interaction capabilities with high-level planning and manufacturing execution system together with a close integration on lower levels enables a direct (re-)configuration and/or direct interaction with shop floor devices/automation systems. The configuration of production cell instances allows both the integration of PLC into the autonomous agent model as well as direct control of actuators and access to sensors that do not require the hard real-time capabilities of a logic controller. The flexibility of this concept, on the one hand, enables the introduction of novel manufacturing control solutions, e.g. based on industrial computers, but on the other hand also allows for integration of legacy systems, e.g. in the form of PLC into the MAS approach.

4.4 Smart Automation of a Flexible Production by means of Evolving Software Agents

Besides the architectural approach of intelligent agents interacting with both, the physical and the digital world in a flexible manner, the evolution of such systems also has to be taken into account. Thus, in the long term it is necessary to go beyond smart algorithms that are embedded into the intelligent entities by introducing an information system approach that provides interface solutions for the physical parts of the manufacturing process, high-level systems and the information model within the agent entities. This information model inside of an agent is an essential part to allow for the learning and evolutionary advancement of its characteristics and functional abilities. An architectural enhancement of the developed approaches that takes into account these desired capabilities is shown in

Figure 4.13 and was inspired by the self-organized and evolvable architecture of Barbosa et al. (2015).

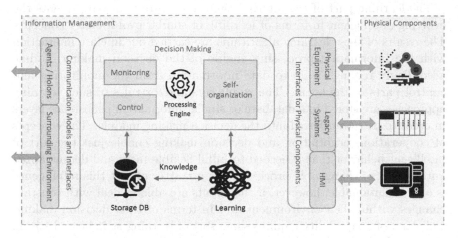

Figure 4.13: Internal information processing and learning architecture of an intelligent agent

The internal agent architecture approach distinguishes between information management part and the physical components that are in close interaction via interfaces. The information management part is further divided into the decision making and information processing part, the heart of the agent. This part is furthermore divided into components that serve interoperability to physical components, legacy systems and HMI. The communication module shown on the left in Figure 4.13 satisfies the required interaction capabilities with other agents and further intelligent components within the direct environment of the manufacturing system.

The interface/physical components modules provides connectivity to the physical devices, legacy systems and display HMI via a Graphical User Interface (GUI) capable of real time monitoring of agent states (Barbosa et al., 2015). The interface to legacy systems provides access to low-level manufacturing execution devices such as PLC and industrial computers.

The decision making module is further divided into several parts that are capable of monitoring and discovering of events and which are able to take control actions while working closely together with the self-organization module of the agent. These sub modules are interconnected though a processing engine that is responsible for managing the information flow inside of the agent and for the triggering of internal actions, e.g. sensing and processing information from the environment, exchanging messages with other agents and supervise the internal behavior of the agent.

The learning part of the internal agent structure finally represents the evolving of an agent in terms of learning or similar evolutionary concepts. The results of these learning algorithms are carried out in close interaction with the information processing engine of the decision making module accordingly relying on up to date information from the monitoring and control parts of the agent. The result sets and/or rules of such learning approaches are stored in a database or similar information persistence layers in order to be retransferred into the decision making module. The concepts of cooperation, negotiation and decision making enable just the sort of intelligent behavior that is needed to fulfill flexible tasks and decentralized automation of changeable processes. As a matter of fact this intelligence is only maintainable, however, if the agents are able to deal with constant changes within their environment and in terms of their decision making. This kind of adaptability requires some sort of learning as indicated within the extended internal architecture of the agent in the beginning of section 4.4. A use-case with regard to an evolvable agent will be presented in chapter 7 in terms of a predictive maintenance scenario.

4.5 Limitations of the Architecture – Communication and Evolutionary Capabilities

In the last two sections an overall architecture for smart MAS solutions had been derived taking into account two main aspects. Firstly, the embedding of intelligent agents into the manufacturing environment in order to represent different resources of a production system has been carried out. Secondly, the evolving of these agents in terms of machine learning based approaches that are able to predict the behavior of resources in a manufacturing environment, e.g. in terms of predictive maintenance or similar scenarios was presented in the next step.

This section focuses on the limitations of current MAS and HMS solutions to realize the architecture in practical applications. These limitations manifest themselves in terms of communication and information modeling abilities and lead to a lack of scalability required for an effective MAS that is well integrated into the shop floor systems if a manufacturing enterprise as well as in high-level systems for mid-term and long-term planning purposes. The lack of scalability in terms of information representation in terms of the agents' internal structure leads to limits in terms agent evolution and learning.

4.5.1 Critical Discussion in terms of Communication Flexibility

The architectural approaches that had been derived in the previous sections form an extensible basis to enable intelligent decision carried out by the agents on different layers of the manufacturing organization. These agents basically cover the required functionalities for planning, execution organization, sensing of their environment, collaboration, negotiation and decentralized manufacturing execution. However, in terms of the state-of-the-art regarding the proposed agent-oriented approaches, the presented architecture lacks several key capabilities that are of critical importance for the scalability of the approaches and for an implementation of the presented solutions into practical use-cases:

- The close integration of the systems on the shop floor opens up high potentials in terms of utilizing up-to-date information in online planning and reconfiguration scenarios. However, the architectures as proposed leave many open questions in terms of the exact representation of information. It is not clear whether the data that comes from field devices such as sensor has to be stored within different data structures than information from automation devices such as PLC.

- The representation of low-level data in terms of a general interaction model is not clearly defined. Specifically, the granularity from low-level system does not match the requirements of higher levels that mostly demand for condensed information. The issue that had been motivated in section 4.1.3 in terms of the information technological requirements cannot not be satisfied in terms of the proposed architectural approaches as the inconsistency of information representation prevents the definition of clear rules in terms of how to aggregate information into valuable KPI or other quantitative measures.

- This lack of consistency regarding information representation furthermore restricts data exchange capabilities with high-level systems. Not all of the decisions regarding production planning and execution can be made solely based on specialized agents, because the agents still lack of certain overall enterprise knowledge. Hence, an information exchange with higher levels is still required. However, without coherent information models, the form of information exchange, its representation and its granularity in terms of interconnecting with higher levels cannot be standardized properly. As a result, certain information exchange methods will still be based on tailored solutions such as customized web services or similar approaches based on non-generic interfaces.

- Accordingly, the formation of SOA infrastructures that fulfill the required flexibility of bringing together all levels of the automation pyramid cannot be realized by means of the proposed multi-agent system design concepts.

To sum up the described issues – the demands for close interaction with the hardware on the shop floor leads to drawbacks in terms of the embedding of agents in a SOA as the integration with higher systems cannot be realized on lower levels, if the information representation and exchange characteristics of data from the field is not modeled in a coherent way. Thus, state-of-the-art approaches are not able to properly address the required recombination of loosely-coupled systems in terms of SOA and tightly-coupled environments such as represented by field buses and other systems on the physical layers of a factory.

4.5.2 Scalability Limitations – Learning and Evolutionary Development

Following the limitations in terms of information modeling and representation capabilities as described in the previous section, the learning and evolutionary abilities of smart software agents are also affected by these issues. The learning of an agent is strongly depending on a contextual model, in which knowledge from the perception of an agent and additional knowledge from external systems can be recombined to a coherent *world view*. Thus, possible learning capabilities of smart agents are equally depending on a proper information technological mapping of the agent in order to store new data that is gathered throughout the production process in an extensible, evolvable context.

The following chapter targets these overall limitations of current multi-agent system solutions by carrying out new form of representing agents in a cyber-space that match the agent's context of the real-world application, in which they are located.

5 Agent OPC UA – Semantic Scalability and Interoperability Architecture for Multi-Agent Systems through OPC UA

The architectural approaches derived in the previous chapter pointed out sine key concepts to enable flexible decision making by means of agent-based, self-organizing and decentralized systems. Means to realize such architecture in practical applications were pointed out despite the limitations that have to faced with regard to current state-of-the-art solutions for the communication in such dynamic networks. Furthermore, a critical discussion of the conventional information exchange methodologies offered some fundamental weaknesses in terms of semantic scalability, extensibility and flexibility in terms communication within MAS.

This chapter aims at targeting these key weaknesses by carrying out a fundamental reorganization of the communication in decentralized systems such as MAS or HMS. For this purpose a connected and integrated approach to the system design of an extended architectural model is considered. This approach intends to be capable of linking high-level design abstractions and principles, such as holons, services, agents, emergence, self-organization and self-adaptation (Leitão et al., 2016) and their according implementation into the system architecture. The realization and harmonization of the required steps in order to reach this overall model is carried out in terms of the design work flow for the realization of agent-based CPS inspired by Farid and Ribeiro (2015). The design principles are visualized in Figure 5.1.

The high-level design principles reflect the desirable system states of engineered manufacturing systems. The according functions are realized by means of the ability to react and adapt the production system to quickly changing environmental conditions by dynamically adjusting the structure or operational mode and gracefully recovering from disturbances (Leitão et al., 2016). Furthermore, the goal of the high-level design principles is to reach a high degree of robustness that ensures an operative system under

© Springer Fachmedien Wiesbaden GmbH, part of Springer Nature 2019
M. Hoffmann, *Smart Agents for the Industry 4.0*,
https://doi.org/10.1007/978-3-658-27742-0_5

Figure 5.1: Design principles in terms of developing agent-based CPS infra-structures, adapted from Farid and Ribeiro (2015)

any circumstances. To realize such stable behavior functional redundancy of the system components as well as highly decoupled control structures are inevitable. The decoupling of control structures allows for a convergent responses that collectively allow the system to explore different actions for present disruptions. (Leitão et al., 2016) Especially the last point can be regarded as transition from high-level planning structure to agent-based approaches.

The high-level design principles had been addressed by the requirements for the desired architecture presented in the beginning of chapter 4. The basic infrastructure of a suitable MAS/HMS-based architecture was derived by extending the existing reference models for MAS/HMS within industrial manufacturing.

The missing parts of an overall infrastructure that fits all the required design principles shown in Figure 5.1 consist in the *CPS system integration*, the realization of *agent behavior in CPS* and the consideration of *domain-specific aspects* of the manufacturing system. These topics will be accordingly addressed in terms of the present chapter and according to the paradigm that "the reference architecture should be as technology and system agnostic as possible to maximize the range of systems in which it can be instantiated" (Leitão et al., 2016). The addressing of these demands regarding a holistic application of the proposed architecture are carried out by applying technical solutions with a highly generic scope such as the OPC UA standard.

Accordingly, the missing links between the different levels of the design principles as well as between the layers of the factory hierarchy as shown by the automation pyramid are targeted by means of serving generic interoperability among the systems. In order to make use of a consistent technology to service the desired interoperability the OPC UA protocol and meta modeling standard is selected for this purpose.

In order to cover the integration of legacy systems in the context of CPS system integration the generic shop floor interoperability standard OPC which is still present in many industrial applications is regarded firstly. The integration of this standard into the flexible architectural solutions of OPC UA is performed on an OPC Classic – OPC UA Bridge that was carried in early works of the author. This subject will be covered in section 5.1.

The next step in deriving a holistic interoperability framework consists in the integration of domain-specific models for automated systems. The model generation as well as integration with the hardware of the manufacturing plant is carried out by modeling solutions that rely on the OPC UA meta model.

The subsequent sections cover the development of the OPC UA enabled MAS framework by elaborating in detail the necessary steps and preconditions as well as the realization of the targeted architecture. These sections constitute the main contribution of the present work to the state-of-the-art regarding the embedding of intelligent agents into automated manufacturing systems.

Due to the information modeling and context-related capabilities of the OPC UA meta modeling approach the IEC 62541 seems as the most suitable basis to integrate all information systems along the automation pyramid. The modeling and representation of intelligent software agents constitutes the next step on the way to this holistic integration attempt. Accordingly, this section describes the required steps to reach a full integration of CPPS on the shop floor with their digital twin counterparts. This proceedings finally attempts to realize an interconnection of low-level automation devices with high-level planning systems allowing for optimization and reconfiguration processes in-the-loop.

At first, the integration of legacy OPC systems is covered in section 5.1. In the next step, the information modeling of CPS/CPPS is examined in terms of the OPC UA meta standard (section 5.2). The section further describes the basic proceedings and modeling blueprints of representing smart entities using OPC UA. The following procedure in carrying out multi-agent system solutions based on the OPC UA meta standard is divided into three steps: (i) Firstly, the general requirements and architectural design decision in terms of enabling an OPC UA based MAS are examined (section 5.3); (ii) In the second step, the representation of an intelligent software agent by means of the OPC UA standard is presented in terms of an OPC UA information model for agents (section 5.4); (iii) in the third step, the skills, abilities and communication capabilities are modeled by means of a companion standard for OPC UA (section 5.5). The way, agent entities as well as their communication abilities are modeled, leads to the derivation of domain-specific information models that are able to extend the scope of intelligent agents in terms of the required capabilities (section 9.1).

5.1 Bridging the Gap to OPC Classic

One integral part of the architectural approach that has been shaped in the previous chapter are capabilities to integrate legacy systems into modern approaches. Thus, one especially important feature in terms of carrying out OPC UA based solutions consists in the incorporation of OPC Classic applications into the OPC UA application context, e.g. by means of suitable wrappers Hannelius et al. (2008).

Although the author of the present work carried out research works dedicated to this backwards compatibility with OPC Classic solutions, the subject will not be covered in this work, since details about the integration of these legacy systems would go beyond the scope of this dissertation that rather focuses on information modeling capabilities of OPC UA. The interested reader might refer e.g. to Hoffmann et al. (2016a).

Although the topics of integrating OPC Classic in terms of OPC DA, OPC A&E and OPC HDA applications in the context OPC UA solutions are not covered in detail here, it should be mentioned that the integration of these systems are of vital importance for an implementation of future-oriented manufacturing configurations into grown automation systems. Thus, the author would like to underline that all modeling solutions that will be derived in the following sections, are fully applicable with regard to a generic integration of OPC Classic solutions by making use of the concepts shown in Hoffmann et al. (2016a).

5.2 Information Modeling and Infrastructure for CPPS

The information modeling approach based on OPC UA is performed according to standardized meta modeling paradigms. This implies that the modeling of all entities can be carried out in the same manner and according to the same semantic background. Accordingly, information from different levels of the manufacturing hierarchy can be described in a consistent way and interpreted in a common context – independently of their origin. In this sense OPC UA opens up the capabilities to make the production data that is generated on the very lowest levels of the manufacturing readable and understandable to arbitrary systems on higher levels which are also capable of the OPC UA standard. This methodology prevents the loss of dense information from the field level as the richness of generated data can be maintained in all its granularity while not restricting the according information to be aggregated for certain services such as KPI in the PLM or similar high-level systems.

In contrast to conventional Internet infrastructures, networks of smart entities such as agents are characterized by open, dynamic architectures that are managed by autonomous instances. The communication between these entities intends not to rely on a fixed, centralized server component that manages all incoming client requests and their according responses. Moreover, in such a web of things as characterized by MAS the communication between arbitrary entities requires a dynamic infrastructure, in which every client instance can also function as a server and vice versa. Thus, the capabilities of the communication infrastructure need to comply with these demands in a suitable way.

Due to the object-oriented modeling capabilities of OPC UA the standard is an appropriate choice to satisfy the requirements of cyber-physical systems such as intelligent software agents in terms of an Internet of Things infrastructure. In OPC UA by nature every client instance is characterized by various dynamic abilities such as demanding for information, triggering services at remote instances or manipulating nodes in the address space of the server. These powerful client functionalities can be utilized in order to extend the communication infrastructure in OPC UA based networks in terms of the requirements of dynamic CPS environments.

Taking into consideration these capabilities a full vertical integration approach based on OPC UA infrastructure solutions can be realized by means of the following steps:

1. Integrate OPC Classic legacy systems on the shop floor into a low-level OPC UA infrastructure, e.g. by making use of the OPC Classic / OPC UA bridge described in section 5.1.

2. Equip manufacturing control and high-level systems with OPC UA connectors by embedding OPC UA clients in the according interface modules.

3. Carry out a suitable representation of resources on the shop floor by means of digital twins, and develop the according OPC UA information model to represent these resources as OPC UA instances.

4. Realize the representation of the resources as agents in order to enable intelligence approaches within the physical resources that are enabled to communicate with arbitrary instances.

5. Finally realize communication between these agents and their represented resources with the OPC UA connector modules on higher levels of the automation hierarchy.

This proceeding does not only realize a multi-agent system with intelligence capabilities on the shop floor, but also enables a seamless communication between all levels at any time. That way, intelligent agents as well as high level planning and execution systems are always empowered with up-to-date information of any granularity. Additionally, the information that is exchanged is represented through a consistent information representation context, thus any piece of information can be read and understood by any of the instances involved.

The key concept to realize this sort of architecture in a real manufacturing environment is to enable a bidirectional communication of the CPS represented resources in the shop floor that are able to communicate with other low-level devices through connectivity solutions and at the same time being able to communicate with higher systems and vice versa. The proposal that is made in this work in order to realize the desired concept is to carry out a formal agent instance that consists of an OPC UA client and OPC UA server instance at the same time. Thus, in this approach we make use of the fact that any physical machine is able to simultaneously function as OPC UA client and server instance. The methodology of this approach is depicted in Figure 5.2.

The figure summarizes the communication and interaction capabilities that become possible by enabling OPC UA connectivity in both directions.

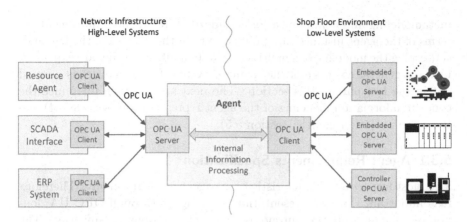

Figure 5.2: Agent instance equipped with OPC UA server and OPC UA client functionalities

The right side of the illustrations represents the low-level systems, e.g. equipped with robots, PLC devices and other manufacturing resources. The left side is characterized by arbitrary instances for a horizontal data exchange (e.g. with other agents) or with higher levels of the production hierarchy. The agent in the middle is located on the edge of these infrastructural environments, being present in *both worlds*. This behavior that becomes possible through the usage of OPC UA addresses precisely the duality of loosely and tightly coupled systems addressed in the motivation and problem description of this work (see chapter 2, section 2.4). Through the low-level functionalities of OPC UA the agent is capable of interacting with tightly coupled systems located on a fieldbus system (right side) and at the same time the agent is part of a service-oriented environment interacting with high-level systems through OPC UA based web services (left side). This systems on each side can be of arbitrary nature as long as they comply with OPC UA in terms of their model representation and communication capabilities.

5.3 Implementation of OPC UA Enabled Multi-Agent Systems

This section describes the requirements that are connected to the desired mapping a fully functional and integrated MAS by means of the OPC UA metamodeling and communication standard. Firstly, the requirements in terms of the agent instances are pointed out. In the next step, the demands concerning the modeling of a multi-agent system infrastructure are explained, before a general coverage of the required system functionalities is pointed out in the last part of this section. The next section then introduce the concrete information modeling of the agent representation (section 5.4) and the communication models (section 5.5).

5.3.1 Agent Requirements Specification

The most important characteristic of an agent is its autonomy. Therefore, it is important that a representation of an agent through OPC UA does not restrict or limit the autonomy of agent decisions in any way. The basic requirements of autonomy is met by three demands that need to be preserved when representing an agent through a meta model standard.

The first demand is an agent's *reactivity*. The reactivity describes the awareness of an agent regarding its environment. As the agent's environment in terms of a manufacturing process consists of real world objects any changes of the real world are noticed by means of sensors or other perception abilities. The agent has direct access to the system it encapsulates or the manufacturing resources it represents, despite the fact that these connections might be constructed of tightly coupled legacy interfaces or through OPC UA connections. In both cases the perception of information from the real world is independent from the algorithms an agent uses to take its autonomous decisions. Thus, the representation of an agent inside of an OPC UA infrastructure does not affect its capabilities of decision-making in any way.

The *proactiveness* of an agent constitutes the second demand of autonomy. In order to enable such proactiveness in a reasonable manner the agent must have knowledge it is able to base its decision on. Traditional agents rather use predefined sets of rules or similar decision tables for this process. However, an evolving knowledge that is needed to enable proactive behavior even in complex situations cannot be covered by static decision models. The representation of agents, however, through an OPC UA information model enables the required scalability of the agent's knowledge model to include learning and the integration of domain-specific ontologies into the conceptual model of an agent.

The third demand of an autonomous agent is its *social ability*. This term summarizes the means of an agent to interact with other entities of the same type, i.e. other agents. In terms OPC UA represented agents this implies the capabilities of agents to communicate with each. Hence, a messaging system needs to be carried out that enables all agents to exchange messages among each other. This ability is also met by the OPC UA based approach and will be covered in section 5.5. As the messaging standard that is presented also includes legacy agents that comply to the FIPA standard the approach is also capable of incorporating grown MAS into the desired architecture.

5.3.2 Multi-Agent System Requirements Specification

In order to enable a representation and interaction of agents through OPC UA not only the agents itself, but also the management components of a MAS have to be mapped by means of an OPC UA representation model. Thus, the essential parts of a functional MAS determined by the FIPA – a Message Transport System, an Agent Management System as well as an optional Directory Facilitator – need to be covered by the capabilities on an OPC UA based agent architecture. The requirements of the elementary components including means to realize them using OPC UA are pointed out in the following.

The *Message Transport System* provides the backbone that enables communication among the entities of a MAS. The traditional transport system such as used in conventional MAS will be entirely replaced by an OPC UA based information exchange mechanism that is further described in section 5.5. The advantage of an OPC UA based message system consists in its scalability. Thus, not only fixed protocol standards, e.g. FIPA compliant messages, can be mapped onto the standard, but also any other standards or context-specific message models can be incorporated into the messaging systems by an extension of the OPC UA information model. Despite the semantic extension of these messages, still all information exchange will be interoperable as a consistent metamodel is used for all representations.

The *Agent Management System* is the central component of a MAS that manages all agents being part of the system. Every agent needs to register with the AMS. This managing instance will be replaced by an OPC UA server that covers all the required demands, e.g. by providing a discovery service for OPC UA based agents. Despite the mapping of existing AMS, the OPC UA based system organization will introduce a number of additional services that facilitate the integration and usage of intelligent software agents.

The *Directory Facilitator* offers yellow page services that enables agents to easily browse for the available capabilities of other agents. The functions of the DF can also be mapped by making use of the OPC UA server instance and its rich services.

5.3.3 Required Functionalities of an OPC UA enabled MAS

One of the main characteristics of OPC UA is its *AddressSpace*. On the one hand, the *AddressSpace* is used to store OPC UA information models including all type definitions of general or domain-specific importance. On the other hand, the *AddressSpace* also manages all instances including their characteristic properties in terms of OPC UA *Objects*. Thus, in order to make use of the high flexibility induced by the *AddressSpace* the communication and representation of the agents is mapped onto *Objects* of the *AddressSpace* in order to make all entities as well as any piece of information globally available.

Another mandatory function of OPC UA consists in the connection methodology of the standard. As it is well known, OPC UA networks always rely on connections between an OPC UA server and an OPC UA client. Thus, all interactions that are realized in an OPC UA based network infrastructure are based on these connection. The basic nomenclature/graphical notation of this connection oriented approach will make use of the conventions illustrated in Figure 5.3.

On the top of the illustration two components that represent some arbitrary components of an OPC UA network infrastructure are connected in terms of a client server connection. Beneath that the general annotation of such components as well as simplified illustrations of the same type of connections are shown. The arrows indicate whether the connection is of general or not of specific character or if the connection represents write or read operations. The following explanations covering the functional requirements of OPC UA enabled agents are carried out by making use of this graphical notation.

5.3.4 Messaging System

The capability for inter-agent information exchange is one of the major characteristics of agents in a MAS. The asynchronous communication enabled by an independent messaging system guarantees the autonomous behavior of each agents. Despite the fact that a sender agent is able to propagate messages at any time, receiving agents can accept these messages,

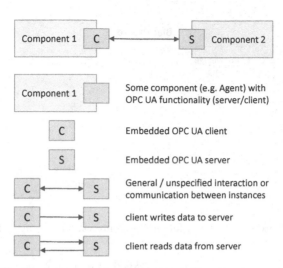

Figure 5.3: Graphical notation of OPC UA based infrastructures

however, these agents cannot be forced to provide feedback or react on the message instantly or at any other time. Instead, every agents has the ability to choose, in which way it will react on the propagated information.

The messages that are propagated within an OPC UA enabled MAS are stored in the *AddressSpace* to ensure a global availability of the information to every instance that might be concerned. Furthermore, the availability of messages can be used for extended supervisory and diagnostics purposes. This methodology of messaging systems realizes an information exchange similar to the ideas of a *blackboard pattern* (Silva et al., 2003), through which the flexibility and accessibility of entities in a network can be increased decisively. An illustration of such message exchange characteristic is shown in Figure 5.4.

Through this pattern, the gathering of information and sharing of knowledge among a high number of entities can be realized with significantly lower efforts than based on direct messages between agents. This is due to the fact that every agent that might be affected by the published information can obtain the according message.

The realization of this pattern in terms of the presented OPC UA based approach is carried out by creating OPC UA object instances for messages that have been sent. The according message objects are instantiated as OPC UA nodes and can be stored in the *AddressSpace*. Due to the general

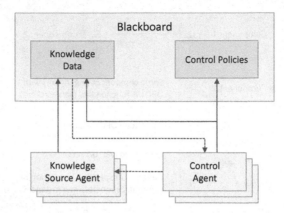

Figure 5.4: Blackboard pattern for multi-agent systems (Dong et al., 2005)

accessibility of the nodes within an OPC UA *AddressSpace*, every agent or other instance is able to read the message and extract its information. As every node that is instantiated into the *AddressSpace* needs to obtain a unique identifier in terms of the *NodeId*, the according identifier is generated including a timestamp as well as a string containing the sender agent and some other randomly generated chars. This way, each message can be identified uniquely and the presence of two nodes with the same *NodeId* is made impossible. The *Message Object nodes* can function as *Attributes* of other nodes in the fashion of the object-oriented modeling methodology of OPC UA.

As all information exchange within OPC UA networks is solely performed in terms of client server connections the communicating parties are obliged to function as an OPC UA server and/or an OPC UA client. In terms of this information exchange methods, three different scenarios for realizing message sending and delivery are conceivable (see Figure 5.5). While the possibilities 1) and 2) realize the communication on direct messages, the system configuration in 3) makes use of a sort mediator to deliver messages to receiving agents.

In order to make use of the messaging system illustrated in 1), all agents are committed to implement a server and a client component as previously described in section 5.2. The implementation of both connector instances makes sure that every agent is able to send and to receive messages throughout the OPC UA based network architecture. In order to send a message, the sending agent's OPC UA client creates a *Session* for establishing a con-

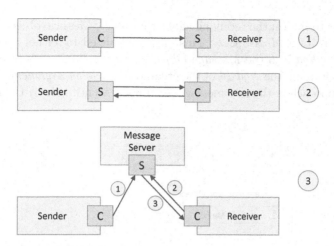

Figure 5.5: Messaging methods using the OPC UA client server paradigm

nection with the OPC UA server of the receiving agent. In the next step, the sending agent creates a message object that complies a predefined message type. The *Call* of the *Send Method* can either use an OPC UA *Method* or can directly write the message into the *AddressSpace* of the receiving server instance:

- In the fist alternative, the OPC UA client of the sending agent calls a *Method* within the *AddressSpace* of the receiving agent's OPC UA server. The payload of the actual message is mapped onto the *Argument* of the *Method* call. This approach is comparatively limited as the *Arguments* of a *Method Call* are not able to carry *Objects* or *References*. Thus, in through this methodology it is not possible to use any semantics of the information model for the message payload.

- For the second methodology, the sending agent's OPC UA client writes data directly into the *AddressSpace* of the receiving agent's OPC UA server. Figure 5.6 shows a sequence diagram that contains the process of a the detailed message exchange in terms of this proceeding. In order to allow for a message exchange in the described way, parts of the receiving agent's *AddressSpace* need to be accessible by all agents in terms of writing rights. To realize a restricted access of other agents, each agent is able to declare a certain node, e.g. a *Folder* object as writable. This way, other agents are able to create *Message Objects* within this folder without controlling the receiving agent. In this scenario the *Send Message* is called in order to inform the receiving agent that a new Message object has been create in its *AddressSpace*. This can be realized by either placing the *Send Method* into the *Message Object* or by sending a separate *Method Call* containing the *NodeId* of the created message as *Argument*. Through both approaches, the new message is uniquely identifiable by the receiving agent.

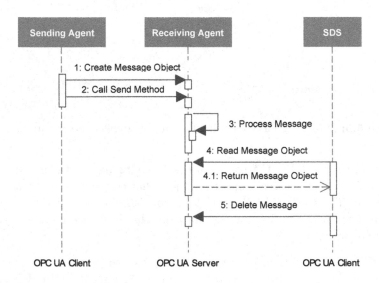

Figure 5.6: Sequence Diagram for the communication between a client sender and a server instance as the receiver

The receiving agent is able to read and process the message from its *AddressSpace* at any time. After processing the message and possibly performing a task that is connected to the message content, the receiving agent might be able to delete the message. However, in some system configurations an agent might not be responsible for this decision, as the message might contain information that could be of higher importance for supervisory or control reasons. In such cases, a higher level control and supervisory system such as a SCADA instance is able to read the messages by using its client functionalities. In the example shown above, the SCADA system can trigger a deletion of the message within the *AddressSpace* of the client. In practical application, this sort of behavior could be realized defining fixed time ranges or a determined number of messages to be stored within an *AddressSpace* of an agent before the messages are automatically deleted. In the meantime, higher level control systems take over the autonomy of this particular *AddressSpace View*.

The second message exchange methodology that can be used according to Figure 5.5 consists of the sending agent represented by an OPC UA server and the receiving agent acting as a server. As through the previous method, all agents need to implement both, a server and a client component, in order to enable read and send message operations in both directions. The sequence diagram of this methodology is shown in Figure 5.7.

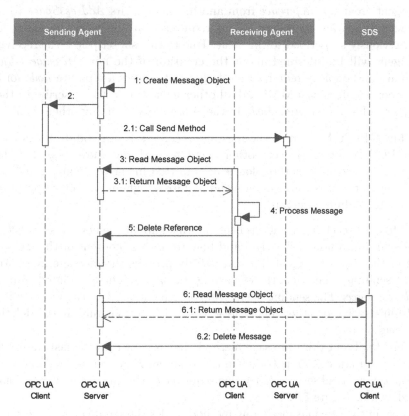

Figure 5.7: Communication between a server as sender and a client as the message receiver

Similar to the previous method, the sending agent creates a *Message Object* based on an existing type definition. However, in contrast to the first method, the sending agent writes the *Message Object* into its own

AddressSpace. In the next step, the sending agent informs the receiver about the new message objects. This can be performed in three different ways:

- The OPC UA client of the sending agent calls a *Send Method* in the *AddressSpace* of the OPC UA server of the receiving agent. This proceeding is visualized in detail in Figure 5.7. The *Argument* of the according *Method* contains the *NodeId* of the created *Message Object* as *Argument*.

- The second approach uses OPC UA *References*. In this case, the sending agent creates a *Reference* from another *node* in its *AddressSpace* to the according *Message Object*. This requires that the receiving agent has previously subscribed to this *node*. Due to this subscription, the receiving agent will be informed about the creation of the new *Message Object*. This methodology requires that each agent creates a separate *node* for all other agents of the MAS and all other agents need to subscribe to their particular *subscription node* in the *AddressSpace* of any other agent.

- The OPC UA client of the message receiving agent could also poll the OPC UA server of every other agent to find new messages. This last and the previous methodologies are omitted in the following as they are comparatively resource intense and require a high number of subscriptions and/or polling intervals.

After the receiving agent has been informed about the new *Message Object* it decides autonomously when and how to use the content of the message. After the receiving agent did successfully process the message, it informs that sending agent that the according message might be deleted from its *AddressSpace*. The sending agent will consequently delete the message if no administrative or other higher level control functions inhibit the deletion process.

The final and third message method that is shown as the last messaging option in Figure 5.5 functions similar to an email system by making use of some central mail server. A detailed sequence of this messaging methodology is shown in Figure 5.8.

Like in the previous messaging methodologies the sending agent creates an object of the *Message* type definition. That instance contains the according payload of the message to be sent to the receiver. However, unlike to the direct communication methods, the message object is stored into the *AddressSpace* of central message server. A *Send Method* is called afterwards from the *AddressSpace* of the message server the message has been sent to. This method creates a *Reference* from the *Inbox Folder Node* of the receiving

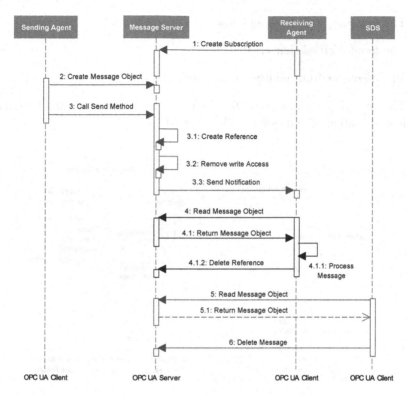

Figure 5.8: Communication between agents using a message server

client to the created *Message* object in the *AddressSpace* of the message server. The *Send Method* also withdraws write access to the *Message* object to prohibit changes of the *Message* object at a later time.

An advantage of the last method is that messages could be sent to multiple agents at the same time, i.e. by creating *References* from the *Inbox Folder Nodes* of all receiving agents to the *Message* object. The methodology of this process is similar to the second message sending method. Due to its complexity a detailed explanation of this proceeding is provided in Figure 5.9). All OPC UA client instances within the receiving agent will be automatically informed by making use of an existing subscription that points to the *Inbox Folder Node* of their own agent's OPC UA server instance. As shown in the Figure, the intend message transfer is realized by four consequent steps:

1. The creation of the message object

2. The *Send Method* that creates

3. all *References* from multiple receiving agents

4. The receiving agent's own OPC UA server instance informs its client instance about the message to be processed.

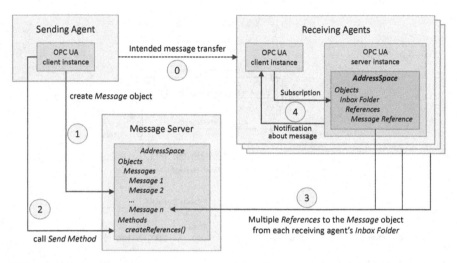

Figure 5.9: Transmission of a message through messaging server

Another advantage of this methodology is that the according server instances of the receiving client do not need to be up and running during the time range the *Message* object is sent. As the *Message* object is stored within the *AddressSpace* of an independent messaging server the receiving client is able to access the message at a later time. However, due to the fact that active subscriptions are only possible on running client and server instances, the according client after creating a *Session* to its server instance has to browse the *Inbox Folder* of the message server in the latter case in order to receive the message.

Similar to the other two message exchange methods the receiving agent is free to process the message whenever he is ready for doing so. The administrative rights regarding the removal of messages from the message server's *AddressSpace* can be given to one of the receiving agents or – with a higher priority – to some higher level system for control monitoring or supervisory functions such as SCADA. Another possibility is to remove the message when all *References* from the receiving agents have been removed by the receiving agent's client instances.

A comparison of the advantages and drawbacks of the elaborated message system realizations in terms of the summary in Table 5.1 leads to a decision in terms of the messaging methodology used for the present framework. Although all of the mentioned message systems have got their advantages and drawbacks, the most elegant solution consists in the third message system that involves a central messaging server.

The first two message systems are based on a direct communication between the sending and receiving agent. The first solution is the leaner approach compared to the second one as it only relies on a single OPC UA *Session*. The performance gain due to this advantage overcompensated the slight drawback that write access within the *AddressSpace* of the receiving agent need to be granted for the message transfer. Thus, the first method is privileged in comparison to the second one and will accordingly be favored.

By establishing a central messaging server instance, the third solutions proposed in Figure 5.5 is connected to many advantages. Firstly, neither the sending nor the receiving agents are obliged to implement an own OPC UA server instance just to exchange messages. Second of all, the central messaging server also increases the availability of the messages to be exchanged and facilitates the addressing of multiple recipients. Another important aspect that distinguishes this approach from the direct message approaches consists in an enabling of the blackboard pattern for information exchange as already mentioned above. Thus, balancing the advantages and drawbacks of all proposed solutions, the third alternative might be the most promising way to realize message exchange based on a scalable, semantic messaging concept in MAS.

One important point, however, has to be carefully taken into account when operating a MAS using the proposed method – the risk of a failing message server in terms of a single point of failure is connected to decisive consequences to the entire MAS. Thus, it is important to make use of the redundancy concepts proposed by OPC UA. Accordingly, the initialization of the message server should be realized in terms of these concepts and by means of at least two independently operable messaging server instances in order to prevent a possible loss of information. Through the usage of a robust redundancy system for the message server, information is very likely not to get lost in practical use-cases. Thus, the central message server with

redundant backup solutions also outperforms the other messaging systems
in terms of data integrity.

Summarizing all mentioned points and finally comparing the messaging
systems, the first and the third solutions from Figure 5.5 could be equally
utilized to realize the message exchange. However, due to the fact that
agent instances intend to run on small embedded devices rather than on
powerful industrial computers, the choice in favor for the lean agent stack
becomes obvious. The messaging solutions that will be pursued in terms
of the framework carried out in this work is characterized by the central
messaging server solution.

5.3.5 Data Storage

Another central point in terms of enabling MAS through OPC UA is to
establish suitable data storage solutions. The data persistence layer within
OPC UA multi-agent systems has to be carried in a different manner than
for traditional MAS, as conventional data storage principles for MAS do
not comply to a standardized ontology. However, in order to server a global
interoperability not only between agents, but also with other systems, the
usage of semantic technologies in terms of ontologies and information models
is demanded. This fundamental difference between OPC UA based MAS
and traditional MAS is pointed out in Figure 5.10.

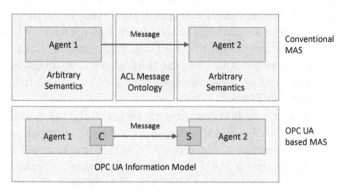

Figure 5.10: Comparison between conventional and OPC UA based MAS in
terms of message semantics

The messages that are exchanged between agents of a traditional MAS
might comply to a certain set of semantics, e.g. the definition of a specialized
FIPA message. Seamless communication between the agents can only be

realized, if all participating entities comply to this standard or if the systems or agents are able to parse a foreign ontology definition. This sort of proprietary standardization works sufficiently for encapsulated systems with a reasonable variety of participating agents. However, as soon as agents from other MAS or entities from other system components intend to take part in the communication with these agents, the mapping of different ontologies becomes complicated is generally not manageable.

By using an OPC UA information modeling approach to address this issue, many interoperability problems can be solved on the spot. By modeling a standardized ontology for the representation of arbitrary information within the production system, not only messages can be represented in a continuous way. It is also possible to store information from different sources in a consistent way. Thus, messages between agent are able to transport genuine production data from the lower shop floor levels to other agents or to higher planning and management systems.

In order to make use of such an approach, the overall data storage for volatile and persistent information has to be realized by means of the OPC UA *AddressSpace* that also contains all information related to the standardized OPC UA information model in use. By making use of globally available *AddressSpaces* in arbitrary places, information could be exchanged seamlessly between agents and other systems simply by writing data into the *AddressSpace* of the according OPC UA server instances.

One particular requirement of this concept with special regard to MAS, however, consists in a fine-grained rights management that is needed to preserve the autonomy of agents. Thus, even in holarchical system configurations like in MAS certain actions of the autonomous entities need to be prohibited in order to avoid mutual restriction of the agents' autonomy by themselves. A feasible rights management could consist of the following aspects:

- *Write access* (create, delete, alter of *Nodes* and *References*) should be granted only to the agent that is the original publisher of the information. In order to realize the management of these rights in a flexible manner, each *Agent* should be assigned to a unique *Agent* object in the globally available *AddressSpace*. This *Object* constitutes the root for all data that belongs to this particular agents. This requirement is considered in the derivation of an OPC UA based agent information model within section 5.4.

- *Read access* should only be grated by the agent the stored information belongs to. In usual operation mode, read access is granted to all parties that might be interested in the generated data, e.g. other agents for cooperation and negotiation or high-level control systems like SCADA. Particular restrictions of the information access should only be performed, e.g. in large scale MAS or if some other proprietary systems are connected through OPC UA that should not get access to certain information.

In terms of the data storage, a further distinction between the different types and usage of information has to be made. In this context the range of information utilization and a possible propagation of the according data have to be taken into account:

- Firstly, there is some sort of *personal* information that only concerns a particular agent or is at least of major relevance to that agent, e.g. its personal perception data or the information of machinery resources the agent is connected to or represents.

- Secondly, another type of information has to be distinguished that is of major relevance for the entire multi-agent system or is at least consumed by a number of agent entities.

Due to the possible coexistence of client and server instances within one physical agent entity, the first type of information should be stored within the *AddressSpace* of the OPC UA server instance of that agent the information concerns. With regard to performance, this approach is favorable in comparison to the storage within a remote *AddressSpace*, because the agent has direct access to its own OPC UA server without the requirement of an established *Session* from its client to another server. Moreover, the agent is able to access its own data under special circumstances, e.g. if the network interaction is disrupted for any reason. However, with regard to significantly important information that agent should pursue an appropriate backup strategy, e.g. by storing certain information on a remote server. This possibility will be discussed in later below.

For the discussed data storage concept within an agent's *AddressSpace* the agent has to implement both, a client and a server instance, which poses certain requirements to the available memory within the embedded device. A feasibility evaluation of typical low-level devices in terms of the needed memory and performance is given in the use-case chapter of this thesis.

Besides to the local storage in the *AddressSpace* of its own server instance, an OPC UA based agent is also able to store information in the *AddressSpace* of a remote server. This remote server can be located on any other physical machine that is part of the OPC UA network infrastructure. If the data transmission protocol supports the underlying protocols such as TCP the remote server could even be instantiated in the form of a cloud server. However, regardless of the location of the OPC UA storage server, an implementation of redundancy concepts is inevitable due to the risk that is connected to a *single point of failure* approach.

Additionally, the management of read, write and general access rights has to be carried out in a structured manner in order to facilitate the organizational efforts connected to the central storage implementation. By making use of the object-oriented information modeling principles of OPC UA this data structuring process can be significantly facilitated. The design decision for a general structuring of the agent's information that was made in terms of this work provides an object instance for each agent being part of the MAS. All information that belongs to a particular agent is accordingly stored as a sub node of the agent's object. That way, every piece of information can be precisely distinguished and the according access rights can be carried out following clear principles.

In practical applications of the central storage concept, each agent that is part of the MAS will automatically trigger the initialization of a new *Agent* object in the *AddressSpace* of the central storage OPC UA server when the agent registers with the MAS for the first time. The registration process is performed by means of *Session* between the according agent client and the OPC UA server instance. As every *Agent* object is unambiguously assigned to a particular agent instance by means of the *NodeId* as a unique identifier, each agent instance will have exactly one information representation in the OPC UA server's *AddressSpace*. In order to exchange data with the server instance, the client of the agent entity has to make use of a *Session* with the central storage server. This *Session* is either maintained after the initial agent registration or might be established every time the client needs to exchange information, such as reading or writing data. In general, i.e. in MAS applications of reasonable scale, sessions between agent clients and data storage servers should be maintained in order to allow for seamless interaction with reasonable reaction times.

One drawback of the central storage solution consists in the usage of higher network resources, because every data related interaction is connected to network traffic. In contrast, if an agent embedded OPC UA server is used for the storage of information, network traffic is only needed, if another agent or connected system requests for particular information from that agent. Despite the higher usage of network resources, however, the usage of the central storage approach is connected to a couple of advantages:

- Each agent *knows* immediately where to find information about and from other agents in the MAS.

- The storage of information from many agents in a central OPC UA *AddressSpace* significantly facilitates the structuring and filtering of information, especially if data from different agents can be linked to each other.

- The server instance can make use of history server concepts as described in the state-of-the-art section by embedding a suitable database in its backend. Thus, complex data filtering and *Queries* can be passed on that database instance, where the data collection and aggregation is performed efficiently.

- Usually, servers for central storage solutions as described above will be located on separate physical machines rather than on embedded devices. Thus, the computation performance will generally be sufficient to handle a higher number of active *Sessions*.

Thus, in terms of realizing MAS applications, both systems – the local agent's *AddressSpace* and the central OPC UA server – will be utilized for storing information. For comparatively low amounts of data that are of particular interest for the agent the data belongs to, the local storage solution will be applied. Also, negotiation processes that require a direct interaction between agents, will preferably make use of the local storage concepts, because only small amounts of information need to be transferred in a frequent manner. However, in configurations that require an exchange or large data sets and an accessibility of this data to a higher number of agents, the central storage solutions will be preferred. Concluding the concepts, a local storage with accessibility when needed underlines the autonomy concept of the agent, whereas the central storage solution offers capabilities for the requirements of large scale systems or the information exchange with external applications.

5.3.6 Reading Values from Remote Devices

As already shown in section 5.2 an integration of field devices into the context of an OPC UA based agent is quiet feasible by incorporating device embedded OPC UA into the agent by means of a client instance (see Figure 5.2).

Aggregating Server for Accessing Field Devices

In terms of multiple data sources in the field the agent's client is able to integrate the values of these remote OPC UA servers in terms of an *aggregating server* (see Figure 5.11). All devices that are connected to the client instance of the agent will be accessible through the OPC UA server of the same agent. This server instance models all remote OPC UA servers from the field in the form of objects that represent the underlying devices on the shop floor. Other clients or agents from the MAS are able to access these devices through the *AddressSpace* of the aggregating server embedded in the agent. If some client requests data from field devices the according client containing the aggregating server will pass through this request to the underlying OPC UA servers in the field. After their response to the agent's client instance, the interested data can be read by the requesting client.

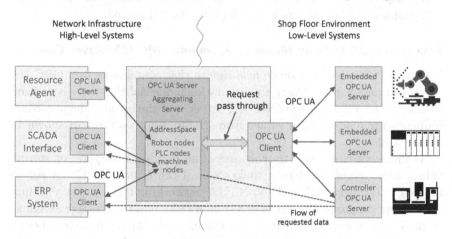

Figure 5.11: Aggregating server concept for passing through information from field devices

If several clients from other agents or third party systems request access to the same device data in the field the agent containing the aggregating server will not establish multiple *Sessions* to the field servers as the requested information can be sufficiently represented within the *AddressSpace* of the aggregating server granting data access to multiple consumers. Due to this methodology, multiple similar requests can be handled without a waste of resources. Furthermore, active sessions will only consume resource, if concrete data requests are being placed to the client containing the aggregating server.

In the case that some interesting variable values from the field might be requested with a high frequency, the agent containing the aggregating server is able to establish a *Subscription* to the according *node* within the field level OPC UA server. After the subscription has been placed, the agent's client will be notified about changes of the variable value every time they occur. Thus, the agent is able to change the according representation of the data variable within the aggregating server's *AddressSpace* ensuring that the provided value is always up-to-date. Consequently, if other agents or third party systems attempt to access the device's value with a high frequency, the agent's aggregating server can respond immediately by offering an up-to-date value from its *AddressSpace*. This, way the communication can be limited to interactions between remote clients and the aggregating server and risks for performance issues between the agent containing the aggregating server and OPC UA servers in the field can be minimized.

Access to Field Data by Means of a Remote OPC UA Server Concept

If an agent is bound to a single field device that contains its own OPC UA server instance, e.g. a PLC or an automation industrial PC, there is no need to implement an additional OPC UA server into the agent. The according sequence for reading data values from these remote servers in shown in Figure 5.12

As the agent is not able to make use of an own *AddressSpace* for storing values from remote devices it makes use of the central OPC UA server located somewhere within the OPC UA networking infrastructure. Thus, the concept proposed here works in a similar fashion as the message exchange using a central messaging server. In the example the agent subscribes to an interesting variable, e.g. of a sensor. Additionally, the agent might read certain values of the field device. After the response of the according OPC UA server in the field, the agent performs a write operation on the central server's *AddressSpace* in order to update the according *node*. After the value has been updated, arbitrary agents are able to access these values in terms of read operations on the *AddressSpace* of the central OPC UA server.

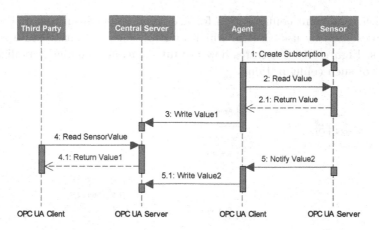

Figure 5.12: Accessing data values from remote devices containing an OPC UA server instance

A drawback of this approach becomes obvious at the bottom of Figure 5.12. As the subscription to values from the agent only activates information exchange, when the sensor data changes, the values are written to the central OPC UA server with a certain delay. Thus, it might be possible that another agent or third party applications access a data value within the central server's *AddressSpace* that is out-of-date. In the case, the according field level value has not changed for a longer time period, third parties are not able to access the actual value, as the value will not be updated by the subscribing agent. Even though, OPC UA does not claim real-time capabilities, the aforementioned sort of delay should be avoided when possible. Also, read operations on any interesting variable should be enabled at any time. Thus, even if an agent connects only to one resource, the implementation of a dedicated server instance within the agent's physical machine should be considered, especially when dealing with important information from the field.

Reading Remote Sensor Values form Other Agents

Sensors are a vital part of each agent in order to enable a perception of their environment. Usually, the according sensors are located in the near periphery of an agent, i.e. the according CPS. However, in some situation it might be suitable that a number of agents share certain information about their environment, e.g. for saving costs due to sensor reduction. Thus,

another important configuration for accessing values from remote devices
is represented by use-cases in which multiple clients share certain sensor
values. Figure 5.13 illustrates how the information exchange is realized in
terms of such configurations.

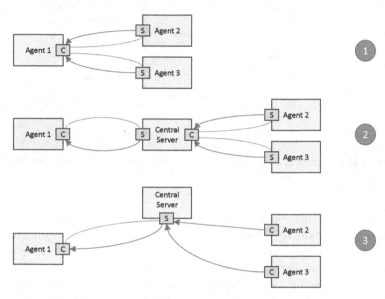

Figure 5.13: Reading sensor values from remote agents

In all of these constellations, Agent 1 attempts to read sensor values form
Agent 2 and 3. All different constellations involve persisting the according
sensor values to an *AddressSpace*.

In terms of the first two approaches the remote agents containing the
sensor perception modules store information within the *AddressSpace* of their
own OPC UA server instance. In approach 1, Agent 1 connects directly
to Agent 2 and Agent 3 by establishing two *Sessions* to the according
agents. In approach 2, an additional (central) OPC UA server is used as
an aggregating server incorporating the data from Agent 2 and 3 in its
AddressSpace. Accordingly, the agent requesting data only needs to establish
a single *Session* to the aggregating server. As this central server can be
requested by multiple agents, the agents providing data in the field only
need to establish one single *Session* each to the aggregating server in order
to publish their information. Assessing these two approaches, the first one
should be preferred, if reading sensor values from remote agents is rather an

exceptional case. The second approach should be chosen, if sensor values are frequently read from remote agents.

Approach 3 shown in Figure 5.13 is realized by a rather proactive behavior of the agents in the field providing sensor values. In this case, the agents establish *Sessions* to a central OPC UA server accordingly pushing sensor values to its *AddressSpace*. This third approach has the advantage that the values within the central OPC UA server, which aggregates all connected agents from the field, are always up-to-date. Furthermore, any agent that attempts to read values from remote agents only need to establish one *Session* to the central server.

5.3.7 White Page Services

In traditional MAS agents are able to browse their environment and to identify other agents through the white page services offered by the agent management system. As there is no overall information model, i.e. an address space, embedded into traditional MAS infrastructures, the agents need to make use of the white page services through messages. In OPC UA based MAS all information that is needed to properly describe and identify an agent instance in the network can be mapped to the OPC UA *AddressSpace*. For accessibility and availability reasons it is convenient to store these sort of global information into the *AddressSpace* of a central OPC UA server similar to the central messaging server or remote value data access server mentioned before. Furthermore, redundancy concepts can be used to carry out multiple instances of the server in order to reduce the risk of failures and offer a high availability at all times. As every agent is able to access the services of this central server instance, their client instances are able to directly trigger the according method calls and there is no need for any message exchange to perform white page services.

The information about an agent's white page entry usually consists of its name and its actual state. At startup, an agent is obliged to register itself using the white page services in order to be browsable and identifiable by other entities in the network. Following to an accurate startup of an agent its state and its name in terms of a unique *NodeId* should be determined. Thus, the information that is exchanged with the central OPC UA server to establish a *Session* is sufficient to create the white page service entry of the according agent. This is due to the *Session* specification, which obligates *Session* objects to contain the name of the OPC UA client. Within the

MAS configuration carried out in this work the OPC UA client name shall be equal to the agent's name in the white page service registry.

Another application of the *Session* service in terms of the white page registry consists in supervision capabilities of the agent's state. The OPC UA server is able to regularly check whether the OPC UA client of an agent is reachable. If the OPC UA server does not receive any acknowledgment following to these requests, the *Session* to the client is terminated, otherwise the state of the agent is shown as active or available. Additionally, the agent itself should be able to change its own state at any time when being connected to the server. If the agent's state is stored within a *Parameter* the agent is able to change the value of the *Attribute* that describes the state *Parameter*.

The data set of an agent's white page entry can be stored into an OPC UA agent *Object* that is part of the proposed agent representation information model described in section 5.4. All agents that are represented by the white page services will be accordingly grouped within an agent *Folder* object that contains *References* to the according agent *Objects*. Any agent is able to obtain a list of those *References* by requesting the according array from the server.

Certain agents of an MAS might be unavailable or characterized by another significant state. The white page services offer means to obtain agents of a certain state. This service can be realized by either of the four following ways:

Query Service The *Query Service* provides enhanced filtering services by passing on incoming requests to an available backend, e.g. a database. If no extended database is available, the OPC UA server is also able to process the query and return a list of agents that are characterized by a certain state.

Method calls If the *Query Service* is not available at the server, a specific *Method* definition located in the *AddressSpace* of the server can perform the requested action by returning the according list of agents.

Agent state Folders The desired functionality can also be fulfilled by means of a reasoned structuring of the *AddressSpace*. In this case, a *Folder* is instantiated for each possible state of an agent. When the state of an agent is updated, the references from the according state folders are also updated by pointing to the agent with the actual state. Thus, a list of

References from a specific folder always contains a list of agents currently characterized by this state.

Client based Filtering While the previous approaches are performed server-side, this method shifts the filtering operation to the client. By means of this approach the client obtains a *Reference* list containing all agents and is subsequently able to browse the state of each agent. As the server-side services are usually connected to higher efficiency, the client-side approach should only be selected, if no other service is available.

5.3.8 Yellow Page Service Realization

Yellow page services provide an overview about available capabilities in a multi-agent system. The traditional MAS concept provides yellow page services by means of the directory facilitator. Similar to the usage of other agent management system services like the white page service, an agent is required to send a message to the DF in order to make use of the yellow page services. Within an OPC UA based MAS a similar service can be used by writing the capabilities of the agents into the according *Agent* objects located in the *AddressSpace* of an OPC UA server. Due to similar advantages as already outlined in the description of the white page services, the yellow page service registry should also be located on a central OPC UA server.

In accordance with the information modeling capabilities of OPC UA complex abilities can be stored as child objects of the according *Agent* node respectively specified by means of the *AgentType* definition. An agent is able to contain a list of capabilities by storing the according *Capability* objects in the *AddressSpace* of the agent representation. Detailed description about the modeling of capabilities is provided in section 5.4.

The proposed concept enables a general availability of requests concerning the yellow page services. Thus, any agent is able to obtain all capabilities that are present in the MAS as well as the according agent. Moreover, an agent should be capable of getting a list of agents with a certain capability. The approaches of obtaining such lists are identical to the approaches for getting a list of agents with a certain state as described in the previous section by

- using the *Query* services,

- calling a *method,*

- using *Folders* for each ability and

- client filtering list of agents.

Unlike the white page services that characterize each agent with a unique state, each agent might have several abilities. Thus, for structural reasons it might be feasible to organize the capabilities of an agent within a *Folder* object that points to the different *Abilities* objects with the *Organizes* reference.

5.3.9 Discovery Process in MAS and Dynamic Networking

One essential part of the startup phase of OPC UA based agents consists in the establishment of a *Session* to a central OPC UA server containing the white page services. In order to connect to the server for getting the endpoint information needed for the *Session* an agent has to know the address of the central server. For convenience reasons the IP addresses of central servers with global services present in the network could be predefined to fixed values. However, this proceeding might be inflexible. Especially, if the agent moves to another MAS or networking environment, the configuration need to be adapted manually. Due to this drawback, a flexible allocation of the central OPC UA server's address should be implemented. There are two main approaches for the realization of such flexible allocation:

Fixed allocation in a subnetwork For dynamic computer networks, the Dynamic Host Configuration Protocol (DHCP) server is used to dynamically allocate entities to available IP addresses. Thus, a DHCP server can be used to assign the network configuration dynamically to an agent. This configuration contains among others an IP address, the subnet mask and an IP addresses of the default and name servers. However, an additional field in which a custom field like the address of a central OPC UA server is not available, thus a DHCP server cannot be used to assign the necessary central servers to agents at startup. Thus, another way of configuring the network could be realized by:

- during the construction phase of a MAS, certain address can be reserved for the central OPC UA servers.

- Depending on the configuration of the MAS a fixed number of IP addresses in a subnet should be reserved for these server, a feasible number in standard configurations could be two addresses.

- As subnetworks are used to divide all available IP addresses of the entire network, fixed consecutive address of a subnetwork can used reserved for these server.

- Thus, it could be specified that two fixed positions in each subnetwork are reserved for central servers, e.g. the third an fourth position.

- This approach is more flexible than other DHCP related proceedings as the agents are able to find OPC UA server, even if the agents change network locations and are moved to another subnetwork.

Local Disovery Service (LDS) with multicast support The LDS with multicast support is a well-defined allocation strategy of the OPC UA specification (IEC 62541-4, 2015). Using this approach, multiple server addresses can be defined in the network environment. This could be of special interest for load balancing reasons or to migrate redundancy concepts for avoiding single point of failure or similar issues.

In conclusion, a multi-server environment based on multicast LDS in combination with a DHCP server should be used to enable dynamic IP address allocation on the one hand and flexible discovery of servers within a subnet on the other hand.

5.3.10 Incorporation of Traditional MAS through Gateway Agents

An important feature of the proposed framework is the integration of legacy systems into the ICT architecture. Thus, in order to guarantee backward compatibility to conventional multi-agent systems, the OPC UA based MAS should be capable og integrating agents or entire MAS derived on conventional approaches, e.g. communicating through the FIPA ACL standard or similar proprietary conventions. Due to missing OPC UA client functionalities, traditional agents are neither able to access the central OPC UA server nor capable of using the OPC UA based messaging system proposed in section 5.3.4. Consequently, a sort gateway is needed. This can be realized by means of a gateway agent that is capable of receiving and sending ACL messages and implement OPC UA client functionalities at the same time. The gateway agent maps an OPC UA information model that is located in the *AddressSpace* of a central or local OPC UA server onto the received ACL message and vice verse (see Figure 5.14). This way the gateway agent is able to integrate legacy agents into the OPC UA based communication (Hoffmann et al., 2016e).

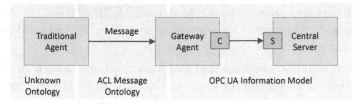

Figure 5.14: Gateway agent and message mediation from conventional agents

From the traditional agent's point of view the gateway agent represents an entire MAS consisting of an AMS, a DF and other agent entities (Figure 5.15). The gateway agent mediates the incoming requests from the different traditional agents to the according OPC UA based services and ensure communication integrity between legacy ontologies and the OPC UA information model. This is performed by parsing the traditional agent message and creating an OPC UA *Message* object that contains the according information. Messages from the traditional agents that are addressed to the virtual AMS or DF cannot be forwarded by means of OPC UA and are processed by an internal logic of the gateway agents. This logic parses FIPA ACL compliant message and extracts the information that is relevant for the AMS and DF service definition, i.e. white page and yellow page service functionalities. The gateway agents then executes read or write operations within the OPC UA *AddressSpace* and is accordingly able to obtain the required information or write necessary data into the *AddressSpace*. Subsequently, the agent gateway is able to answer of the traditional by wrapping the obtained information into an ACL compliant format.

The virtual entities of the gateway agent need to have a unique name, whereas the address of all virtual entities is identical. The according address always points to the socket of the gateway agent. This way, the gateway agent of the OPC UA based MAS will receive all incoming requests from traditional agents. When requested by means of white page services, the gateway agents always returns its socket address, which represents the address of the former traditional AMS. As traditional MAS only contain one socket address – the one of the – AMS requesting to this address does not imply any differences for the traditional agent. This procedures has the advantage that traditional agents do not need to be reconfigured prior to an incorporation into OPC UA based MAS and will function plug & play.

At the bottom of Figure 5.15 the forwarding of incoming messages from traditional agents is shown. This concept is used for the registration of

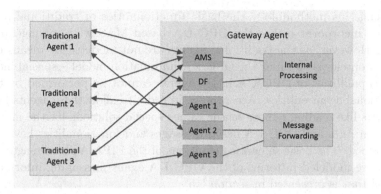

Figure 5.15: Gateway agent imitating functionalities of traditional MAS

traditional agents by means of white page services. In OPC UA based MAS the registration of a new agent is performed by establishing a *Session* between an OPC UA server and the according client. In order to integrate traditional agent entities into the OPC UA based MAS as seamlessly as possible, the gateway agent creates virtual representations of these agents as shown in Figure 5.16. These virtual agents are equipped with OPC UA client that are able to make the registration request at the central OPC UA server. The identity of the traditional agents that is present in the registration ACL message can be used to create OPC UA agent *Objects* of the same name. In Figure 5.16 the *Objects Agent 1*, *Agent 2* and *Agent 3* are created in order to map their identity to the according traditional agent entities with the same identifier.

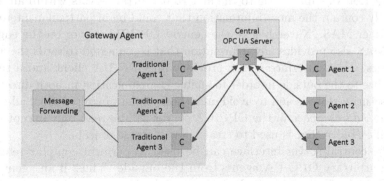

Figure 5.16: Registration of traditional agents using the gateway agent concept

Using this methodology the basic functionalities of traditional agents can be incorporated into the OPC UA based MAS with manageable efforts. However, as the traditional agents are not modeled by means of an overall ontology – such as the OPC UA information model – special interaction capabilities of OPC UA based agents cannot be performed by these traditional agent entities. Accordingly, in order to allow these agents for operations like reading remote sensor values special solutions for the mapping between ACL and OPC UA based messages and functionalities have to be performed. The basic semantic definitions of the FIPA ACL messages, however, are modeled in terms of the OPC UA agent messaging information model that is presented in section 5.5.

5.3.11 Interconnection of Multiple MAS by means of Mediation Agents

Finally, the last part enabling MAS infrastructures through an OPC UA based network consists in the interconnection of multiple MAS. This functionality can be of special interest in large scale systems that cover multiple application domains or use-cases and are accordingly separated into different MAS. There are two basic configurations among multiple MAS that need to be covered separately: (i) Firstly, two or more MAS coexist in parallel existence and and equipped with equal rights; (ii) The second possibility is represented by the case that one MAS might be embedded within another MAS.

The interconnection of two coexisting MAS is depicted in Figure 5.17. The central OPC UA server of each MAS represents the interface. This convention was made due to the fact that central servers within an MAS usually contain the most information that could be of particular interested for other MAS. Nevertheless, the central OPC UA server on the edge of the MAS also provides any information that is requested towards the agent entities. Each interface server provides an OPC UA client and a server instance to the outside in order to enable communication in both directions. For security reasons and to avoid unauthorized data manipulation, all nodes in the *AddressSpace* of the OPC UA server will be read-only except for a specific *Folder* that is used to create new message objects.

The central server interface can be used to perform complex requests from arbitrary OPC UA agents from the outside. Thus, if an agent of a MAS attempts to read a value from a *node* that is stored in the context of another MAS the agent will connect to the central server. In order to

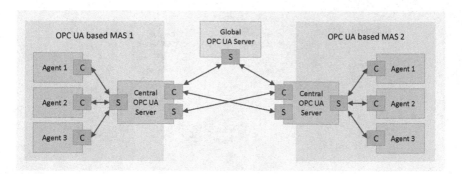

Figure 5.17: Interconnection of two coexisting MAS

be able to process all incoming requests the central server incorporates the *AddressSpaces* of all agent instances in terms of an aggregating server. That way, the central server at the edge of the MAS is able to access all information that is stored in the agent instances or in the near periphery of these agents.

The global OPC UA server that is shown in the middle of Figure 5.17 is an optional entity and can be used for a discovery among multiple MAS. Thus, this global server provides white page services for the detection and identification of all central servers representing the entry point of different MAS. Due to the scalability behavior of OPC UA the different MAS itself can be regarded as abstract agent entities being part of a global MAS. Thus, the information model that will be derived in section 5.4 is also suitable for describing an aggregation of MAS in a global context. Due to compatibility reasons it is highly recommended that each coexisting MAS should define its own *Namespace*, even if the inner entities represent similar objects. This way, the uniqueness of *NodeIds* can be guaranteed even in the context of global agent interaction between multiple MAS.

The second main approaches mentioned above in terms of combining multiple MAS consists in an embedding of one MAS into another. The concept and realization of this approach is shown in Figure 5.18. In this configuration, the embedded MAS is encapsulated in terms of a single agent. Accordingly, an interface agent – the *Mediation Agent* – will represent the single point of contact to this particular MAS by providing a client interface. From the internal as well as from the external point of view, the mediation agent will can be regarded as a standard OPC UA based agent.

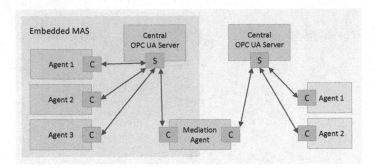

Figure 5.18: Incorporation of an embedded MAS

If an agent of the surrounding MAS attempts to send a message to the embedded MAS that contains a request for action, the mediation agent searches for agents with the requested abilities by means of the yellow page services. If suitable agents can be found, the central server within the embedded MAS forwards the message from the mediation agent to the according agent instance. If more than one agent of the embedded MAS is able to perform the requested action the mediation agent will select the most suitable or *best* agent. Means of how to select *best* agents will be further regarded in the use-case chapter of this work.

5.4 Representation of Intelligent Agents by means of OPC UA

After the basic services and functional requirements that are imposed to MAS have been realized by making use of native OPC UA services and server client functionalities the next steps consist in the actual information modeling of the agent entity and of the semantic context, in which messages between agents are embedded.

Thus, in this section, the representational model of an agent is carried out by extending the OPC UA metamodel through a custom ontology stack. The goal of this semantic information modeling stack is to realize the agent model representation that was outlined in Figure 5.2 and represents an agent that bridges the gap between physical resources and their digital twin accordingly coexisting in both worlds on the shop floor levels. This information modeling of the agent finally intends to enable intelligent software agents *to speak* OPC UA in order to communicate holistically in MAS environments that are designed in terms of the metamodeling standard. These aimed abilities induce several requirements in terms of the agent model to be carried out:

- In order to communicate using the OPC UA metamodeling standard, an agent needs to represent all of his knowledge by means of an OPC UA information model.

- As the knowledge of an agent comprises of several elements such as his perception, his world view and other inbound information, the entire representation of an agent including his identity as well as all information flows need to be placed into the same modeling context.

- The representation model of an agent needs to resemble in a consistent manner in order to integrate the knowledge of different agents into one overall context.

In addition to these CPS related requirements the OPC UA standard is also able to cope with the needs of a global infrastructure in terms of an IoT. Such environment is defined as an information infrastructure, to which all objects of the physical or information world are connected to. These objects should be capable of collecting data from other objects or exchange information with any other object. This behavior requires a global availability of all information present within the infrastructure. A consistent implementation of OPC UA as the basic representation form for all agents precisely enables these demands.

Due to the various requirements mentioned above, the design decision taken in this work is to model the entire agent entity in terms of its digital presence in the form of a CPS instance as well as its communication by means of the object-oriented capabilities of the OPC UA standard and its consistent information modeling approaches. The model extension of such OPC UA compliant agent representation and standard are covered in the following sections.

5.4.1 OPC UA AddressSpace Representation for Smart Agents

The object-oriented modeling paradigm of the OPC UA *AddressSpace* metamodel offers various capabilities in terms of designing and representing information in large scale systems. Similar to the definition of *classes* in typical object-oriented programming language the *AddressSpace* concepts offers means to create type blueprints for the instantiation of physical and/or digital objects. The modeling proceedings of these definitions can either occur in a hierarchical manner as well as tailored to holarchical environments such as multi-agent systems. Through this concept, type definitions can be carried out that contain blueprints or assembly instructions for certain objects. The object-oriented modeling allows for an integration of custom type definitions with variables, sub ordinary objects and other existing or custom type definitions in order to model the intended classes.

Thus, the first step to implement agents through OPC UA concepts is the definition of an *AgentType* and its according integration into the semantic scalability concept of OPC UA. This *AgentType* definition consists of all means and capabilities to implement local data storage concepts, e.g. for perception information or data processing, into the agent instance by means of an extended *AddressSpace* model. The organization of this semantic model extension is outlined in Figure 5.19.

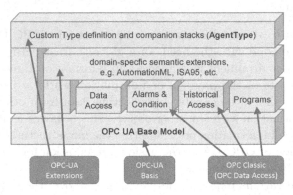

Figure 5.19: Extension of the OPC UA metamodel using the *AgentType*

As already pointed out in the state of the art section the basic setup of the OPC UA metamodel allows for a flexible scalability of the information model that is represented within the *AddressSpace*. Thus, an arbitrary number of node type definitions can be added to the reference model in the form of blueprints for the instantiation of objects or custom variables. These node definitions are summarized in the form of companion specifications, which deliver the semantic context a certain application domains in terms of fully integrated information models. By making use of this concept, the *AgentType* model definition serves as a companion standard that delivers the object-oriented representation of the basic agent within a multi-agent system. In practical applications, these companion specifications are introduced into the *AddressSpace* of an OPC UA server instance by means of an additional *Namespace*.

5.4.2 OPC UA AgentType Model

The proposed information model to represent agents within the OPC UA metamodel consists of a complex type definition. The basic structure of the *AgentType* model definition and its integrability in large scale or multiple MAS is depicted in Figure 5.20.

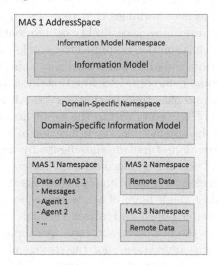

Figure 5.20: Structure of the *AgentType* model stack

The figure illustrates the organization of the according information models with OPC UA *Namespaces*. As an agent can be part of a large scale system or interconnecting with agents from remote MAS, the *AddressSpace* needs to manage the according model instances in a structured manner. On the top, the *Namespace* for the embedding of general information models, i.e. to define the basic *AgentType* definitions for the according MAS, is provided. Beneath these models, another *Namespace* is located containing domain-specific information models, e.g. to enable an integration of domain-specific knowledge with the agents of a specific MAS. The conceptual organization follows the presumption that a dedicated multi-agent system intends to pursue target goals of a certain application domain.

At the bottom of Figure 5.20 the parts of the MAS *AddressSpace* are located that carry the actual information as well as instances of the participating agents. The agent instances located in this MAS *Namespace* contain the actual *Agent Objects* as well as all specifications of the agent's characteristics, e.g. the state of an agent, its occupation or other variables that are directly correlated to a concrete agent instance. In this part of the *AddressSpace* all agents or other instances belonging to the MAS are organized in terms of an object-oriented structure. The remote data shown on the right might belong to another MAS that the current MAS is exchanging information with. Thus, the *AddressSpace* of a MAS is also capable of carrying information from other systems or does at least provides *References* to the interesting objects or variables that represent the remote data.

In order to realize a seamless accessibility of information throughout the entire MAS, the modeling solution is applied to a single *AddressSpace* located at a central point within the MAS infrastructure. This procedure facilitates the realization of some common MAS functionalities such as white page services which require a central (registration) server. Although solutions with many detached OPC UA servers within the agent instances with separate *AddressSpaces* would be also feasible, the utilization of a single *AddressSpace* by linking all information stored in the clients by means of *References* facilitates finding and organizing information. This way both, the autonomy and functional requirements of individually acting agents can be maintained and at the same time, a single point of access for requesting information can be realized. This *pointer* approach realized through OPC UA *References* significantly increases the flexibility and interoperability of a MAS.

The usage of a single *AddressSpace* for the storage of information further requires a unique identification of each entity, variable or other OPC UA typed object that is located in the *AddressSpace*. By making use of the *Namespace* concept in combination with the OPC UA *NodeId*, it is possible to provide an unambiguous access to every node in the *AddressSpace*. By providing a unique number and name for every agent instance, the *NodeId*, which is constructed through concatenation of the *Namespace index* and a unique object name, becomes an unambiguous identifier for each agent. By providing an independent *Namespace* for each MAS, unique identification can be guaranteed among large-scale system infrastructures.

Derivation of the AgentType Node Definition

The construction of the actual *AgentType* node definition is performed by means of object-oriented techniques. One basic paradigm that enables a coherent information modeling for complex model definitions is the usage

of composed objects. These compositions are characterized by *Objects* or
Variables that depend on some other parent *Object*. If the parent object
is deleted, the according sub nodes that are depending on this composed
object, should also be deleted. In terms of an agent type definition, a parent
object might be the perception module object. Sub nodes of this parent
might be concrete instances for sensors or variables that reflect the recent
sensing history of the agent. Thus, if the perception module might be
removed from an agent instance, because it is not be needed for a specific
use-case, the sensor objects should also be removed subsequently. The
OPC UA metamodel for information modeling provides means to realize
this (hierarchically) composed objects. The composition of parent and
depending objects is realized in terms of special OPC UA *References*, i.e.
the *HasComponent ReferenceType* might be used for this purpose.

In order to further group the OPC UA nodes in terms of a hierarchical
organization of information inside the *AddressSpace*, two different concepts
can be applied (see Figure 5.21):

- The predefined *FolderType* allows for the instantiation of a *Folder* object
  that enables the grouping of an indefinite number of nodes. The creation
  of node groups can be performed arbitrarily. However, the different
  objects that are present within a certain group should be characterized
  by a distinctive similarity or even rely on the same type definition. The
  example on the left side of Figure 5.21 describes the grouping of different
  abilities the agent is capable of. The *Abilities Folder* is connected to the
  Agent instance in terms of a *HasComponent Reference*. The individual
  Ability instances are components of the *Abilities Folder* instance.

- The second organization methodology entirely depends on specialized
  References describing the relation of nodes to each other. In the example
  shown on the right side of Figure 5.21, a specialized *ReferenceType*
  definition is carried out by means of the *HasAbility Reference*. This type
  definition that inherits the basic properties of the general *ReferenceType*
  node expresses the characteristic abilities that can be performed by agents.
  This approach provides a higher flexibility in terms of the information
  modeling process at it is not based on the definition of static folder
  objects. However, the scalability and adaptability of this approach is
  quiet limited, because any different sub object type would require the
  definition of a specialized *ReferenceType*. Due to the complexity of this
  approach especially for the modeling of detailed object definitions such as

the *AgentType*, this method is omitted in the following modeling process, unless its usage is connected to significant advantages.

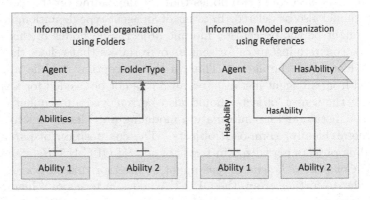

Figure 5.21: Methodologies for grouping nodes within the *AddressSpace*

Thus, in terms of the proposed information model containing the definition of the *AgentType* both grouping methods are used for the mapping of relations between depending objects. In terms of object compositions, the organization through *Folders* is preferred, whereas special types of *References* are used to group independent objects. Following this terminology, an object is called independent, if its existence does not rely on the presence of another node.

The *AgentType* information model is shown in Figure 5.22. The *AgentType* node definition is modeled as sub type of the OPC UA *BaseObjectType* that constitutes the main entry point into the OPC UA *AddressSpace*. Accordingly, the *AgentType* inherits the main characteristics of OPC UA object type nodes and any instantiated *Agent* object will be characterized by its basic properties. The *AgentType* definition is further characterized by several *HasComponent References* to the *Abilities Folder*, the *Sensors Folder*, the *Inbox* as well as to a *State Variable*. The *Ability*, *Sensor* and *Message* objects grouped within these folders are instantiated based on their type definitions shown on the bottom of Figure 5.22.

The mandatory *State* variable is used to store the current status of the agent. As the white page services for the access and registration of agents rely on up-to-date information about the agent's status, the *State* has to be properly instantiated and defined for any agent object present in the MAS. The *DataType* of the *State* is defined by making use of an *Enumeration* that contains an array of Strings listing all available states of an agent,

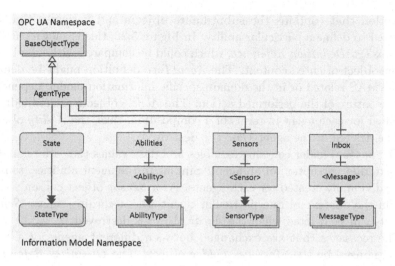

Figure 5.22: *AgentType* definition using OPC UA information models

such as *ready, working, error* or other application/domain specific states characteristics. The *StateType* definition that is used for the instantiation of *State* variables is a subtype of the OPC UA *BaseDataVariableType*. For the development and extension of the *AgentType* information model in terms of domain specific requirements, the *StateType* definition might be further extended by additional variables or objects. An example for such extension could consist in the definition of a properties object that contains additional meta data about the agent that could be of interest for certain processes. In this case, the enhancement the *AgentType* is performed in terms of horizontal extension, e.g. the properties object is added to the *AgentType* definition on the same hierarchical level as the *State* variable. Besides this kind of extension, the definition of further subtypes underneath the *AgentType* definition should be avoided because of the general meaning of the *State* in the context of a MAS. Thus, the definition of the *StateType* should be coherent throughout all agents of a determined MAS.

The mandatory *Abilities Folder* of an agent contains a list of abilities the agent is capable of. The *Ability* objects are used by the yellow page services to propagate the available capabilities of a MAS. The *Ability Folder* might contain an unlimited number of references that point to the according abilities, including zero for the case that an agent is not capable of any action. Each ability is initiated by means of the *AbilityType* definition that contains the subordinate objects and variables that are required to define a particular ability. In Figure 5.22, this is shown through the *HasTypeDefinition Reference*, which could be compared with inheritance in the object-oriented context. The *AgentType* definition might be defined in the MAS related or in the domain-specific information model, depending on the nature of the performed action. The *Ability* objects are individually initiated for each agent in terms of a composition, thus, the *Ability* objects will be deleted, if the agent *Object* is destroyed.

The *Sensors* folder contains variables and data values that are related to the external perception of an agent. Similar to the agent abilities, sensors are individually created for each agents. As a *Sensor* object can only exist within the context of one particular agent, the according sensor *Sensor* objects will be deleted when the *Agent* object is destroyed.

The messages that are exchanged between different agents of a MAS are organized by the *Messages Folder* object. The *Organizes References* shown in the figure are pointed to the *Message* objects the according agent should receive. This special type of reference is defined in IEC 62541-3 (2015) and refers to a hierarchical reference that preserve the independence of a *Message* object. Each *Message* object is derived by means of specific *MessageType* definition that will be further examined in section 5.5 that focuses on the semantic integration of the agent communication.

Subtypes for the Integration of Legacy Systems

In order to preserve the integration of legacy agents into the MAS the *AgentType* has to be further specified by two subordinate type definition that describe an agent by means the OPC UA based approach (*AgentWith-ServerType*) and the *TraditionalAgentType* representing conventional agent implementation (see Figure 5.23).

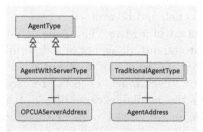

Figure 5.23: *AgentType* subtypes

The *AgentWithServerType* constitutes the class definition for agents that might contain an embedded OPC UA server that can be used in terms of the above described messaging capabilities of to publish sensors data. In this case, the central OPC UA server of the MAS has to be aware of the agent OPC UA server. Thus, the *AgentWithServerType* definition contains the mandatory *OPCUAServerAddress* variable that contains the Uniform Resource Locator (URL) of the local OPC UA server. The server address variable is a subtype of the *BaseDataVariableType* of the basic OPC UA information model. A concrete *OPCUAServerAddress* variable is bound to instantiated agent object and will be deleted if the agent object is destroyed.

The *TraditionalAgent* to the MAS by means of gateway agents. In order to communicate with a specific gateway agent, the traditional agent instance needs to be aware of the gateway agent's socket address. The mandatory *AgentAddress* variable contains this information and is an essential part of conventional agents cooperating in OPC UA based MAS.

Hierarchical Structuring of Agents Being Part of an OPC UA based MAS

Agents are modeled as independent objects. Thus, the embedding of *Agent* objects into the hierarchical structure of the *AddressSpace* is not depending on the instantiation of any other objects. Figure 5.24 shows the proposed hierarchical structure for embedding agents into the *AddressSpace* of a MAS.

The shown information model is divided into three sections. The upper section represents the *Namespace* containing OPC UA basic nodes. The *Namespace* shown in the middle represents the parts of the MAS information model that has been derived in this work. Finally, the lower section of the *AddressSpace* shows the *Namespace* of a specific MAS containing concrete instances of *Agent* objects.

The OPC UA base specification defines two entry points into the *AddressSpace* – the *Types* folder and the *Objects* folder. The *Objects* folder contains all instances that have been created on the basis of the available type definition, including the *Server* object for a self-representation of the OPC UA server. Instantiated objects can be structured as direct sub objects of the *Objects* folder, however a further structuring in terms of additional folders is recommended. The example shown in Figure 5.24 proposes an *Agents* folder to keep all instantiated agents in a consistent place.

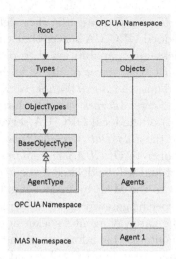

Figure 5.24: Hierarchical structuring of *Agent* objects in a MAS

Instantiation of Agent Objects

The concrete instantiation of an agent is connected to the creates of all additional nodes that are part of the *AgentType* definition and are accordingly required for a consistent representation of the agent. An example for the instantiation of an agent instance is shown Figure 5.25 that contains the full model of concrete traditional agent object. The *AgentAddress* variable that is defined as a mandatory *Variable* of the *TraditionalAgentType* is shown on left. Furthermore, the *TraditionalAgent1* instance contains other *Folder* object and further sub objects that are specified the *AgentType* definition. Thus, the created instance does not only create the variables and objects of its own type definition, but also all required objects of the according parent node types. Hence, the *State, Abilities, Sensors* and *Inbox Folders* have been derived by inheriting the required type definitions from the *AgentType* in the fashion of object-oriented paradigms. The *Ability1* and *Ability2* as well as the *Sensor1* and *Sensor2* object were automatically derived from the exiting *AbilityType* and *SensorType* definitions.

During the initialization phase of the agent, access rights and restrictions are set. Despite the agent no other entity is allowed to create, alter or delete nodes within the *Agent* object. However, the *State* variable and the *Inbox* folder might be altered by the server that sets the *Attribute* field of the *State* variable when the *Session* between the central OPC UA server and the agent

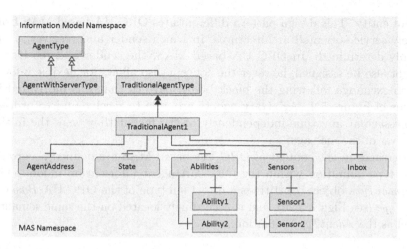

Figure 5.25: Instantiation of an agent based on the *TraditionalAgentType*

is deleted. The OPC UA server is also able to add *References* to the *Inbox* Folder of the agent in terms of delivering incoming messages by pointing to the according *Message* objects. Agents are not allowed to change their name, to alter variables that have been set during instantiation such as the *AgentAddress* or to alter *Message* objects (integrity). However, agents are allowed to remove *Message References* in the case they autonomously decide not to take a requested action (autonomy).

5.5 Semantic Integration of Agent Communication and Interoperability

Following to the general representation of OPC UA based agents and MAS in the context of an object-oriented modeling concept, this section focus on the semantic integration of agents in terms of message semantics, interaction and negotiation capabilities.

5.5.1 The MessageType Object Definition

As pointed out in section 5.3.4, messages enable an asynchronous communication between the entities within a MAS. The whole concept of the messaging system in OPC UA based MAS relies on the fact that *Messages* are characterized as independent *Objects* not belonging to a determined agent entity. This design pattern differentiates OPC UA based MAS from solely service-oriented architectures, in which sender and receiver(s) are firmly determined. In OPC UA based MAS, the sender and receiver(s) might also be assigned, however the concept also allows for flexible information exchange following the black board pattern that was motivated in terms of Figure 5.4. Arbitrary agents are able to receive and to process a message at any time independently of the fact it they were the initial receiver of the message.

Messages in the context of OPC UA based MAS are defined according to a type definition within the OPC UA agent information model. The *MessageType* object constitutes a direct subtype of the OPC UA *BaseObjectType* (see Figure 5.26), and is accordingly located on the same semantic level as the *AgentType* definition.

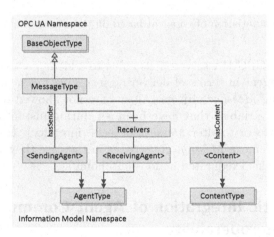

Figure 5.26: Instantiation of an agent based on the *TraditionalAgentType*

Each message that is created for agent-based communication is an instance of the *MessageType* object. Every *Message* object is comprised of exactly one sender and at least one or more receivers. The obligation of a receiver agent is indicated by the *hasSender Reference* pointing from the *MessageType* to the *SendingAgent* instance. The *hasSender ReferenceType* was specifically carried out in terms of the OPC UA based *MessageType* and is a subtype of the *Organizes ReferenceType*. The *hasSender Reference* indicates that the agent it is pointing to, is characterized as a sending agent. A OPC UA *Message* object is only allowed to have one *hasSender Reference* as messages of multiple senders are prohibited.

Besides the *SendingAgent* each *Message* object is also characterized by a *Receivers* folder that organizes the receiving agents of a message. As the *Receivers* folder is a fixed part of the *Message* object, the relationship between the *Message* and the *Receivers* folder is characterized by a *HasComponent Reference*. The *Receivers* folder points to several agent entities using *Organizes References* indicating that the according agents are characterized as the recipients of the message. In contrast to the *hasSender Reference* the usage of a general *Organizes Reference* is sufficient to indicate of the agents as *ReceiverAgents*, because the source of the *Reference* is unambiguously defined by the *Receivers* folder object.

It shall be mentioned that the flexible way, agents are assigned to their roles reflect the adaptability and flexibility of agent interaction in an OPC UA based MAS. Agents are designed with the purpose of constant adaptation rather than with an aim to determine fixed roles to the agents. Hence, each agent that is instantiated in terms of an OPC UA based MAS, is characterized by unique properties that arise from its various interactions with other agents or objects.

The *MessageType* is further characterized by the *Content* object representing the actual payload of the message. The *Content* object is instantiated by means of the *ContentType* definition and accordingly inherits all objects and variables from its super type. The *ContentType* object definition reflects the general representation of message content such as text, but might be further specified in terms of subtypes such as specialized *ContentType* definitions, e.g. a *GetNextProductionStepContentType* might be feasible to characterize a message payload that indicates needed production steps for a product to be finished. These type definitions can be either located in companion specifications or within domain-specific information models that are precisely designed for the purpose of certain manufacturing domain.

Similar to the *hasSender* reference, the *Content* of a message is characterized by a *hasContent Reference* that was defined for the OPC UA based MAS information model. However, in contrast to the sender reference, the *hasContent Reference* is derived from the *HasComponent* reference, because the content is an integral part of the message and can not be separated from its according *Message* object. Each message is characterized by a single *Content* object.

5.5.2 Hierarchical Organization of Message Objects

The hierarchical structure of the *MessageType* and its according *Message* objects is similar to the structuring with regard to the *AgentType*. The embedding of *Message* objects into the *AddressSpace* of an OPC UA application is shown in Figure 5.27. Both, the *MessageType* definition as well as the *Messages* folder are located in the *Information Model Namespace* that also contains all other general type definition components that are related to the OPC UA based MAS. The single *Message* objects as instance of the *MessageType*, e.g. *Message A*, which are organized by the *Messages* folder, are located in the MAS *Namespace*.

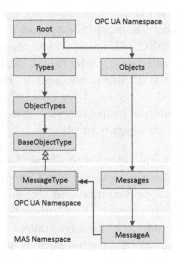

Figure 5.27: Hierarchical structuring of *Agent* objects in a MAS

With regard to the general conventions of OPC UA based information modeling, existing *Message* objects might not be altered after instantiation, because each *Message* object within an OPC UA *AddressSpace* needs to be clearly defined by a *MessageType*. In order to introduce new message features or additional domain-specific knowledge into the model definition of a message, the *MessageType* definition has to be extended through optional or mandatory *nodes* or subtypes that might be defined within a domain-specific information model or the internal information model of the MAS. This procedure ensures that all messages that are sent within the MAS are understandable by all agents that might be interested in the message content.

5.5.3 Compatibility of OPC UA to Legacy ACL Messages

This section focuses on two key aspects: (i) Firstly, the needed backwards compatibility of derived OPC UA *Message* objects to legacy message semantics of the FIPA ACL standard is shown; (ii) Secondly, the flexibility of the design approach is demonstrated by the modeling of the ACL compliant type definition based on the *MessageType*.

The information modeling capabilities of OPC UA allow for an extensive enhancement of predefined types. In order to integrate the compatibility of the OPC UA based agents with messages from ACL based legacy systems, the *MessageType*, the *AgentType* as well as the *ContentType* have to be expanded with regard to the requirements of the FIPA standard. The according information model including the mapping of OPC UA *nodes* to the according message components is shown in Figure 5.28.

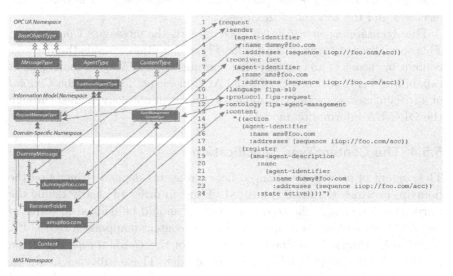

Figure 5.28: Hierarchical structuring of *Agent* objects in a MAS

The sender and the receiver parameters of the message can be mapped by means of the according *MessageType* definitions as shown above. The *TraditionalAgentType* extends the *AgentType* definition to map the requirements of a legacy agent. The according FIPA ACL message is constructed in terms of a parameter list being part of a request, wherein the single parameters are *sender, receiver, language, protocol, ontology* and *content*.

The language parameter describes the agent dialect that is used by the according FIPA ACL message. The protocol parameter further specifies the purpose of the ACL message, i.e. in the case shown above the message intends to send a *request*. In OPC UA based MAS this protocol is mapped by the introduction of the *RequestMessageType* that represents a subtype of the *MessageType* definition. As different protocols might define different parameters, it is necessary to define an additional subtype of the *MessageType* for each protocol that might be used for the message exchange. The available protocols, i.e. message purposes might vary according to the present application domain. Thus, the the specialized *MessageType* definitions to map these protocols are located in the *Domain-Specific Namespace*.

The *DummyMessage* that is placed within the *MAS Namespace* represents an actual instance of the message object. The according elements of the message point to the different parts of an ACL message, i.e. the *sender*, the *receivers* and the *content* of the message.

The *AgentManagementContentType* extends the message's *ContentType*. This specialized subtype of the *ContentType* definition extends the message content by means of an ontology that contains the FIPA agent management capabilities being part of a traditional ACL message. This way, the *Content* of a specific message will be compliant to the agent ontology defined through the OPC UA information model.

5.5.4 The ContentType Specification

The content of messages exchanged between agents depend strongly on the domain, in which the MAS is located. Thus, in order to clearly define the purpose of a message, the according content should be defined in terms of an OPC UA information model. All possible content mappings for OPC UA based MAS should be defined as subtypes of the *ContentType* definition provided by the basic MAS information model. These subtypes might be specified in the form of further subtypes depending on the complexity of the application domain. As the derivation of an information model and its according components is a complex task, a generic *ContentType* definition is provided in this context to facilitate the generation of specialized contents. Figure 5.29 shows the underlying approach based on semantic message definitions that are stored in OPC UA *Objects* and *Variables*.

The *GenericContentType* can be regarded as a blueprint that enables the development of new content types suitable to the needs of an application. The *Variables* and *Objects* provide both, information and semantics for

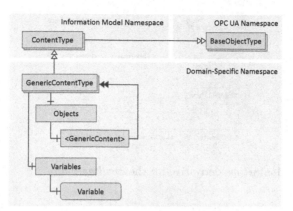

Figure 5.29: The *GenericContentType* for agent messages

the message definition. The *GenericContentType* intends to facilitate the modeling of message contents for the mapping of different message purposes, without the need to carry out a separate *ContentType* subtype for each ACL ontology. Thus, instead of defining new *ContentType* subtypes for every possible message content, the semantics of the according ontology can be mapped in terms of object and variable *nodes* that provide the required content. Detailed examples about *GenericContentType* messages and the derivation of complex *ContentType* subtypes can be found to Appendix B.2.1.

5.5.5 The AbilityType Specification

Abilities represent the core capabilities of agents. By making use of the yellow page services, agents are able to find other agents with certain abilities. In order to register an agent ability with the yellow page services, an *Ability* object is created and published to the yellow page server. The *AbilityType* node definition is a direct subtype of the *BaseObjectType*. The *AbilityType* allows for subtyping to create specialized abilities, e.g. as shown in Figure 5.30.

During the instantiation of an ability object, the variables such as *MinDiameter* might be assigned. Thus, it the *MinDiameter* is set e.g. to $d_{min} = 4\,mm$, the yellow page service won't return the instance of this particular drilling machine, if another agents requests for a *drilling* manufacturing step with a diameter of $2\,mm$.

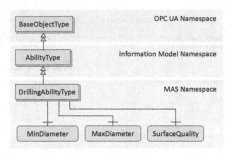

Figure 5.30: *AbilityType* derivative for the *DrillingAbility*

5.5.6 SensorType Definitions

In terms of automation systems, a high variety of sensors is used. Despite the fact that many sensor vendors provide customized OPC UA information models for their sensors, the creation of specific *SensorType* definitions is also possible. The modeling of these sensors is carried out in the fashion of the *AbilityType*, e.g. in terms of customized object and variable *nodes* that reflect the properties of the sensor.

In order to facilitate the identification and modeling of suitable sensors, standardized lists of electrical properties can be utilized, e.g. provided by DIN IEC 61987-11 (2010). The sensors types that are listed in these sort of specifications are structured in a hierarchical way. Thus, the development of customized *SensorTypes* in the *AddressSpace*, e.g. of a domain-specific information model, could follow these sort of structuring.

Existing approaches for the standardization of sensor, i.e. in companion specifications of OPC UA might also be used. A combination of these preexisting type definition with well-structured standardizations and additional specifications/metrics form a powerful tool to incorporate automation environments into a digital context that allows for interoperability from various, heterogeneous sources.

5.5.7 Overall Information Model for OPC UA Based MAS

The overall OPC UA information model for the representation of agents as intelligent entities and their communication through OPC UA based messages is finally shown in Figure .

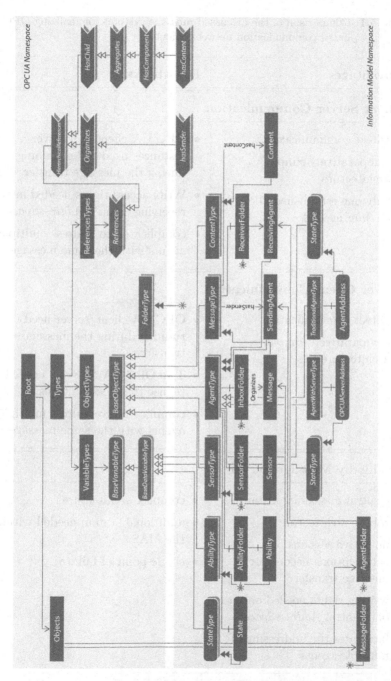

Figure 5.31: OPC UA information model for multi-agent systems

Table 5.1: Comparison of the discussed message systems for realizing OPC UA based communication between agents

Advantages	Drawbacks
Client Server Communication	
• Direct communication • comparatively simple architecture • only one established OPC UA *Session* needed	• OPC UA client and server instances need to be running during the message transfer • Write access rights needed in the receiving agent's *AddressSpace* • complicated to address multiple agents with the same message
Server Client Communication	
• Direct communication • comparatively simple architecture	• OPC UA client/server need to be running during the message transfer • Two OPC UA *Sessions* needed to realize the message transfer • complicated to address multiple agents with the same message
Decidedly Message Server	
• Central messaging instance • robust architecture • no active sessions or subscriptions needed for message transfer • writing rights needed only to one central *AddressSpace* • facilitates the addressing of multiple agents	• complex architecture • additional system needed within the MAS • single point of failure

6 Management System Integration of OPC UA based MAS

The incorporation and mapping of multi-agent systems by means of the OPC UA standard opens up various potentials with regard to a flexible production automation. One major benefit of the extended ICT lies in the interconnection of intelligent software agents with high-level planning systems in terms of vertical integration. These way, the advantages of applying both, centralized planning approaches within top-level systems and at the same time decentralized reconfiguration, can be exploited in the automated production environment.

This chapter shows the potential of the derived approach of OPC UA based MAS by bridging the semantic gap between the functionalities of an ERP system and the capabilities of smart agents on the field level. This high-level system incorporation is performed by means of an open source ERP system that seamlessly exchanges information with the agents of an underlying MAS by means of an *ERP gateway* or *interface* agent.

Section 6.1 focuses on the architectural extension of the existing OPC UA based MAS approach in terms of incorporating functionalities of resource planning systems. Section 6.2 describes the realization of an agent interconnection to these systems by providing the semantic model definitions for an OPC UA ERP information model. This description provides the required *AddressSpace* definitions as well as means that allow for a seamless communication between smart agents in the field and ERP services.

6.1 Architecture Extension for Decentralized Planning

The goal of this section is to demonstrate the benefits of an architecture, in which decentralized autonomous agents are able to interact with *knowledge sources* such as resource planning systems in order to obtain general process information. The basic assumption in terms of this approach is that decentralized agents *do not need* to store all information that could eventually be

© Springer Fachmedien Wiesbaden GmbH, part of Springer Nature 2019
M. Hoffmann, *Smart Agents for the Industry 4.0*,
https://doi.org/10.1007/978-3-658-27742-0_6

of importance for the production process. Instead, an interconnection with high-level systems allows for a flexible access to these information at any time. This way, information of general importance, e.g. the sequences of manufacturing steps or product orders, can be managed by a centralized instance, although all agents are able to use high-level system functions and services whenever they need to.

6.1.1 General Capabilities/Requirements for Resource Planning

The basic capabilities of resource planning and similar high-level systems is to provide a mapping between customer orders and the underlying production resources of a factory. Generally, ERP systems are characterized by tasks to support existing or emerging processes within enterprises in order to rationalize process chains (Osterhage, 2014). The focus of ERP systems lies on the following key competencies:

- Availability of resources,

- delivery reliability and on-time customer service,

- flexibility to react on changing market conditions and

- reduction of cycle times and lowering of costs.

In the manufacturing domain, especially the flexibility of production process is of major importance. Special focus with regard to demanding customers is hereby on the "lot-size 1" paradigm (Osterhage, 2014) that intends to enable the manufacturing of high variant products with reasonable efforts and costs.

In order to realize such dynamic flexibility in the production process information from various sources is needed. On the one hand, information from the market/from the customer side has to be integrated into the ERP system. On the other, up-to-date information from the production also has to be incorporated in the planning process. Figure 6.1 shows typical information flows of modern resource planning systems.

The data sources of the resource planning system shown at the top primarily provide a detailed planning of the production process as well as additional information with regard to the manufacturing products such as work plans, lists of parts and components as well as Computer-Aided Design (CAD) drawings. From the low-level systems of the manufacturing, production data in terms of BDE systems and machine-related data by

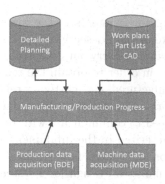

Figure 6.1: Data acquisition capabilities of typical ERP systems (Osterhage, 2014)

means of MDE systems are provided. However, as already outlined in the state-of-the-art section, the information that is delivered from these sort of systems lacks the required granularity of production information, such as sensor data or PLC I/O values. Moreover, these systems are usually only able to provide information in the form of condensed data or selected diagnostics information for SCADA systems. This information, however, is not sufficient to realize reconfiguration processes and reactive planning adaptation based on up-to-date information from the field. One possible approach to overcome this lack of generic interoperability from the low-level to high-level systems is the integration of OPC UA based communication between the shop-floor and high-level planning systems such as ERP.

6.1.2 The ERP Agent as Gateway to High-Level Planning Systems

The extension of OPC UA based MAS by connection modules for high-level resources planning systems pursues the central goal of enabling a seamless cooperation with resource planning systems taking into account up-to-date production data from the factory floor. In this context, ERP systems are capable of managing specific tasks such as customer order administration and production planning in terms of component management and task organization (resource allocation). In any way, the flexibility of these approaches can improved by providing high-granular production related information such as sensor data or detailed error states of machines in the field to to the planning systems.

The concepts of using smart agents in the manufacturing are not new. As described in the state-of-the-art section, many solutions exist, in which multi-agent systems attempt to carry out decentralized planning solutions on the shop floor. However, in order to obtain the required information about the products to be manufactured, e.g. customer related information such as urgent product orders or sequences for the manufacturing of new products, the agents in the field need to communicate with high-level systems in some way. However, as characterized in the fundamental concept description of this work, communication and control flow are usually of unilateral behavior, thus the control flow is managed by high-level systems and only very limited data from the field flows back. Information from low-level systems to high-level application is mostly realized through special interface solutions such as customized adapters and are carried out as shown in Figure 6.2 (Hoffmann et al., 2016b).

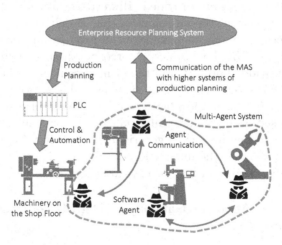

Figure 6.2: Conventional interoperability approaches between MAS and high-level systems

The visualization shows a MAS in the fashion of a cyber-physical systems approach. The agents on the shop floor act as *digital twins* incorporating the according production resources in the form of a digital representation. The agents communicate with each other by means of a certain protocol definition, e.g. the FIPA ACL standard or some other proprietary format. The communication with high-level systems, however, is limited by means a single customized interface. Such interface could be realized by means of a service bus or other proprietary web service definition. Such an approach enables communication between the MAS and e.g. an ERP system, however, the interaction capabilities carried out by such an approach are limited in terms of scalability and ontology extensions.

Thus, in order to increase the scalability of such an approach, the interoperability of MAS with high-level systems such as MES or ERP systems need to carried out in a generic manner. The OPC UA information modeling approach for the mapping of agents and inter-agent communication with the OPC UA standard as carried out in chapter 5 enables precisely such an approach by introducing a generic information modeling and protocol definition for all communication flows between various entities in a factory. The visualization of such generic approach is shown in Figure 6.3 (adapted from Hoffmann et al. (2016b)).

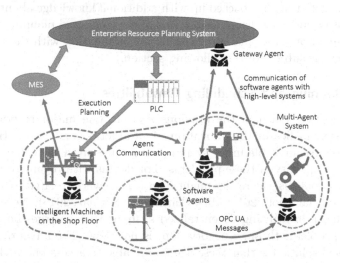

Figure 6.3: Interoperability of OPC UA based agents with high-level planning systems

As in the previous example, the agent represent machines and other production resources in the form of smart entities. Agents are also able to communicate with each. However, in contradiction to the configuration shown in Figure 6.2, each single agent shown in Figure 6.3 is also capable of communicating decidedly with high-level planning systems. These communication capabilities are due to the usage of the generic OPC UA communication standard. In terms of the ERP system, this direct communication of single agents with high-level systems is realized by means of an interface agent, the *ERP Gateway Agent*. This gateway agent is modeled just like the agents in the field, however located in the context of the ERP system application, thus able to establish an internal information exchange with the ERP.

Thus, even if the ERP is not capable of communicating through OPC UA, the gateway that is located on the edge of the ERP and represents an OPC UA client, enables an internal communication with the ERP through a custom ERP interface and a communication with agents in the field through OPC UA. By enabling this communication, smart agents on the shop floor gain access to general production information such as work plans, part, lists and detailed planning data. In this way, the planning and scheduling capabilities can be carried autonomously by the agents through negotiation, but backed up with additional knowledge about market conditions or customer demands in terms of the connected planning systems. How these autonomous planning actions in combination with the ERP are carried out, is subject of the following section.

6.1.3 Planning and Scheduling Capabilities

As already mentioned in the previous section, one major responsibility of ERP system lies in the allocation of production resources. However, centralized resource planning systems alone are often not able to cope with the complexity of these problems. For details with regard to resource allocation problems, the reader might refer e.g. to works related to *Job Shop Scheduling* (Malakooti, 2014; Mirshekarian and Šormaz, 2016). Nevertheless, the paradigm of utilizing decentralized systems that emerge the solution to these problems through negotiation, is not new. The advantage of an agent-based solution is that these agents are able to access knowledge from the planning systems and at the same time are also capable of sensing their environment as well as communicating with other agents. This way, problems with regard to resource allocation can be solved cooperatively and during the process, without the need to anticipate all possible manufacturing configurations in advance (see Figure 6.4).

The gateway agent – which in the Figure 6.4 is referred to as ERP – allows for an incorporation of the ERPs functions into the MAS as naturally as in the communication between agents in the field. The simplistic sequence diagram shown in Figure 6.4 describes how the communication process is carried out in practice. An agent entity is able to subscribe to *orders* in terms of an OPC UA *MonitoredDataItem* subscription. Accordingly, if a

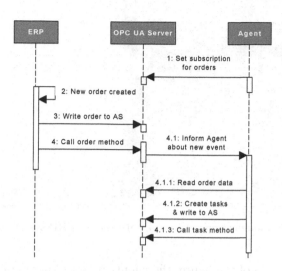

Figure 6.4: Agent subscription to incoming orders

new order is created in the ERP system the agent will be informed as soon as the ERP system publishes the order into the OPC UA *AddressSpace*. This *AddressSpace* can be located either within a server instance that is part of the ERP gateway agent or on a centralized OPC UA server located in the same network as the other agents.

The negotiation mechanisms in terms of a dynamic in-process production planning and reorganization will be covered in the use-case section of this work. By means of the use-case, negotiation and production scheduling techniques will be combined with a bidding mechanism that will be introduced in the next chapter on agent machine learning.

6.1.4 Decentralized Organization Based on a Blackboard Approach

The idea behind a cooperative planning in terms of the ERP system in combination with smart agents is inspired by the blackboard pattern shown in section 5.3.4. In this context, the gateway agent on the edge of the ERP system can be interpreted as some sort of *knowledge source agent* as shown in Figure 5.4. After pushing the according information to the blackboard, every other agent is able to use this knowledge. The according architecture with regard to the MAS with incorporated ERP system is shown in Figure 6.5.

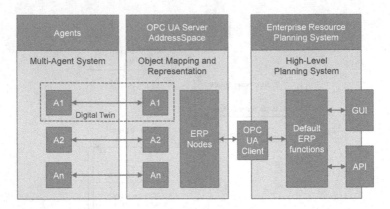

Figure 6.5: Communication between agents and an ERP system interface

On the left side of the figure, the agents representing manufacturing resources on the shop floor are visualized. The *AddressSpace* – the blackboard – is located within a central OPC UA server that provides virtual mappings of the agent instances in the form of objects. The interconnection between the agents in the field and their representation in the server's *AddressSpace* actually constitute a digital twin being part in both world, the manufacturing shop floor and the virtual MAS mapped by the OPC UA server. The *AddressSpace* also contains the nodes that are required for the mapping of ERP functions such as production sequences and other resource planning related information. The OPC UA client on the edge of the ERP acts in the same way as the agents in the field, however sharing information from the ERP system. Possible GUI and API features of the ERP function as usual.

6.2 Incorporation of an ERP System Into OPC UA Based MAS

The mapping of ERP functions and capabilities into the context of an OPC UA based MAS requires the development of an OPC UA information model that contains the according node type definition for resource planning functions. This section provides the basic node definition needed for such an ERP companion specification. The functionality of the ERP mapping is shown in terms of incorporating an open source ERP system with an OPC UA based MAS. In the last part of this section the benefits of such an approach are explained.

6.2.1 The ERP Information Model Specification

The *AddressSpace* is extended based on the existing type definitions of the OPC UA base information model. The extended *AddressSpace* including agent objects and the according ERP extensions is shown in Figure 6.6.

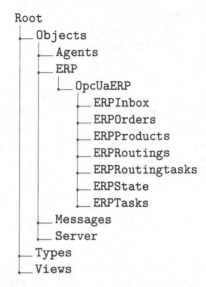

```
Root
  └─ Objects
       └─ Agents
       └─ ERP
            └─ OpcUaERP
                 └─ ERPInbox
                 └─ ERPOrders
                 └─ ERPProducts
                 └─ ERPRoutings
                 └─ ERPRoutingtasks
                 └─ ERPState
                 └─ ERPTasks
       └─ Messages
       └─ Server
  └─ Types
  └─ Views
```

Figure 6.6: Tree view of the ERP *AddressSpace*

The ERP object is located directly under the Objects folder that organizes all instances of an OPC UA *AddressSpace*. It contains all possible ERP system wrappers, e.g. the *OpcUaERP* object. The *OpcUaERP* objects contains further sub nodes that are necessary to map basic ERP functions onto the OPC UA *AddressSpace*, e.g. an *ERPInbox* for message exchange with agents, the *ERPOrders* folder for storing customer orders. The *ERP-Products*, *ERPRoutings* and *ERPRoutingtasks* represent the mapping of production plans, manufacturing sequences and possible production steps for the finishing of a customer order. The *ERPState* provides information about the general status of the ERP, the *ERPTasks* are used to organize the tasks that assigned to the agents of the MAS.

The ERP orders folders contains customer orders including basic properties such as desired delivery date, total grand, detailed item descriptions and order dates, the current status of the order and its priority. The hierarchical structure of the orders folder is shown on the left in Figure 6.7.

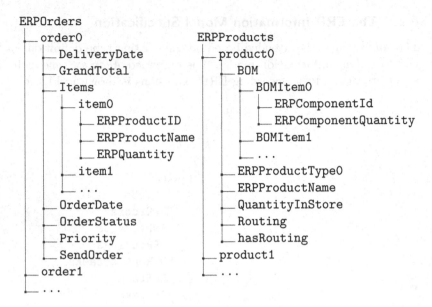

Figure 6.7: Hierarchical structure of ERP orders and products folder

The product order further contains multiple items that are connected to the according order. Items are characterized by the *ERPProductID*, *ERPProductName* and *ERPQuantity* of the order's item. The products itself are organized in terms of the *ERPProducts* folder. Products are composed of material lists which are also referred to as bill of materials (BOM). This lists are further characterized by different components that are part of the product to be manufactured. The product is additionally characterized by a *ERPProductType*, *ERPProductName* and the *QuantityInStore* which contains possible products in stock. The *Routing* of a product provides sequences of the manufacturing process. A detailed description of the product's *route* through the production is provided in Figure 6.8.

The routing of a product contains the associated routing tasks that are ordered in the same sequence as the manufacturing of the product needs to take place. Every associated task contains a *RoutingItemName*, a *SequenceNumber* and a flag that describes whether the task is active. The *RoutingTaskId* represents a reference to the routing task that need to be carried out to fulfill the associated task.

The routing task itself contains the actual manufacturing step that can be carried out by means of manufacturing resources (see right side of

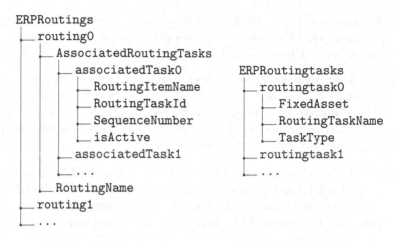

Figure 6.8: Hierarchical structure of the ERP routings and routing tasks

Figure 6.8). The *TaskType* of a routing task represents the according machine capability that is needed to carry out the task, e.g. *milling* or *drilling*. The *RoutingTaskName* is just an identifier for the task and the fixed asset contains immobile parts of the factory that are directly correlated the pursuing of the according manufacturing step.

By making use of this information model and the according type definitions various order, products and tasks that are organized within ERP systems can be mapped to the *AddressSpace* of an OPC UA server. The agents in the field that represent the actual assets of the factory, are able to access this data, e.g. routings and to use this knowledge in terms of their scheduling, planning and negotiation techniques. In terms of a proof concept with regard to this functionality, the next step consists in the implementation of an ERP system that enables a functional incorporation into an OPC UA based MAS.

6.2.2 Incorporation of an Open Source ERP System

The concepts that are enabled by the ERP *AddressSpace* extension in terms of an OPC UA information model are suitable for the integration into practical applications. In the course of this integration process, a detailed list of open source ERP solutions has been examined in order to find a suitable system architecture that is easily integrable with the existing framework components. This list is shown in the appendix B.2.4 in table B.1.

In order to pick a suitable ERP system that fits in with the existing software stacks of the derived framework, a number of requirements had been carried out for the selection:

- The software in use should be open source, i.e. without further restrictions in terms of licensing models for the usage in a framework ecosystem. This requirement is carried out in accordance with the philosophy of the intended framework that should be implementable without barriers regarding proprietary software solutions.

- As all other components of the OPC UA based MAS framework are carried out by making use of Java-based software stack, the ERP should also make use of Java. This way, the software components, e.g. an agent and the ERP system API can be easily integrated in terms of an internal information exchange.

- The ERP software should be actively managed and maintained by the developers. This is of special importance for Java-based software in order to preserve interoperability with other software components written in Java.

- The internal software stack of the open source ERP system should be easily accessible in terms of code customization. This requirement is of special importance to analyze the conditions to incorporate internal ERP knowledge into the OPC UA *AddressSpace*.

A ERP software systems that fits most of the stated requirements comparatively good is the *OFBiz*® open source ERP system, which is a top-level project of the Apache Software Foundation. This software was selected for further developments.

6.2.3 Cooperative Reactive Production Planning

The interconnection of the open source ERP system with an MAS in the field supports smart agents in their independent decision-making for autonomous production planning and scheduling. The functionalities and knowledge of the ERP system is solely represented by means of an OPC UA *AddressSpace* mapping of the information that is important for the agents. The cooperative and reactive planning capabilities that are enabled by means of this knowledge are shown in terms of a practical use-case that is described in section 8.2, in which the autonomous planning and execution of a manufacturing use-case is examined.

7 Flexible Manufacturing based on Autonomous, Decentralized Systems

7.1 Learning Agents for Flexible Manufacturing

In multi-agent systems applications, different types of agents can be distinguished, such as simple reflex agents, utility based agents and learning agents (Russell et al., 2010). Current applications of MAS are often characterized by simple reflex agents that exhibit a library of predefined rule sets. These agents have a determined behavior and are comparatively easy to set up, however they are also characterized by limited intelligence. Especially, dynamic and highly-complex environments such as manufacturing systems would require a large set of predefined actions in order to adequately respond to changing boundary conditions, such as defect machines or other unexpected behaviors: "The complexity of many tasks arising in domains, including robotics, distributed control, telecommunications, and economics, makes them difficult with preprogrammed agent behaviors. The agents must instead discover a solution on their own, using [adaptive] learning" (Busoniu et al., 2008). Due to this complexity demands the learning capabilities of agents play a vital role in applications of the manufacturing domains, such as for scheduling, quality and process control (Jedrzejowicz, 2011; Rocha et al., 2014; Cristalli et al., 2013). By making use of learning strategies, agents capable of adapting their behavior to new environments and learn new actions in order to improve their performance over time. There is only little generally accepted definition or common understanding of learning agents. However, in additions to the general model of evolving agents (see Figure 4.13) the design model shown in Figure 7.1 illustrates commonly used design patterns of a learning agents based on Russell et al. (2010).

According to this model, an agent is split into four parts: the learning element is responsible for improving the performance of the agent, the performance element determines which actions shall be taken based on current

© Springer Fachmedien Wiesbaden GmbH, part of Springer Nature 2019
M. Hoffmann, *Smart Agents for the Industry 4.0*,
https://doi.org/10.1007/978-3-658-27742-0_7

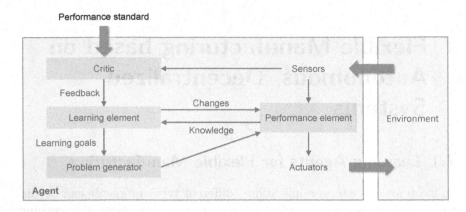

Figure 7.1: Learning agent according to Russell et al. (2010)

perception, the critics provides feedback to the agent and the problem generator confronts the agent with unfamiliar situations. In Russell et al. (2010), the key components of this architecture are explained in terms of a school analogy of a student and a teacher: The student takes an exam at school (critic), the teacher corrects the test and provides instructions about possible improvements (feedback). In this context, the teacher corresponds to the learning element, wherein the student can be regarded as the performance element. Finally, the problem generator is characterized as the teacher suggesting a new experiment to the student that is added to the students knowledge base.

The implementation of a learning agent for a specific application is connected to the design of three key aspects, i.e. the *knowledge representation*, *learning algorithm* and the *training set* that is used by the learning algorithm (Sardinha et al., 2005). The knowledge representation can be regarded as a function that maps perception to actions, while updating an internal state $F : S \rightarrow A$. In most situation of real-word applications, however, it is infeasible to strive for a perfect mapping, i.e. representation of knowledge, due to a large or infinite number of possible states. Therefore, the problem is usually reduced by means of a function that approximately fits to the agents learning algorithm.

In order to illustrate the connection between knowledge representation (performance element) and the learning algorithm (learning element) of an agent, suitable domain-specific examples can be carried out. In terms of manufacturing use-cases, the knowledge representation of an agent can be mapped in terms of a price function, e.g. for manufacturing costs, that maps a prediction for future prices (Sardinha et al., 2005):

$$\text{PredictedPrice}_t = \alpha \cdot \text{Price}_{t-1} + (1 - \alpha) \cdot \text{PredictedPrice}_{t-1} \qquad (7.1)$$

Equation 7.1 can be regarded as the knowledge representation of an agent, wherein the knowledge is represented by means of the α coefficient. The coefficient itself is determined by an agent learning algorithm based on the Least Means Squares (LMS) method that can be classified as a supervised machine learning technique:

$$\alpha_t = \alpha_{t-1} + \beta \cdot (\text{Price}_{t-1} - \text{PredictedPrice}_{t-1}) \qquad (7.2)$$

where β represents a predefined learning rate. In this context, the agent is making use of its current perception of the environment (Price_{t-1}) in order to estimate the α coefficient, while adjusting the knowledge representation to achieve a maximum performance. Other knowledge representation methods go beyond the described exponential smoothing estimation, e.g. by making use of linear weighted functions, collection of rules or Artificial Neural Network (ANN) (Sardinha et al., 2005).

Among the various learning strategies that can be applied in order to pursue algorithm like the price function mentioned above, such as incremental learning or batch learning, one of the most promising approaches in the current state-of-the-art are represented by ML techniques.

7.2 Machine Learning in Learning Agent Environments

Machine Learning can be considered as a discipline, targeting the design and development of algorithms that extract patterns from empirical data. In terms of ML approaches that are applicable for manufacturing use-cases, three major learning strategies can be distinguished – supervised learning, unsupervised learning and reinforcement learning (Monostori, 2003). Despite the selection of a strategies, agents are capable of applying either of these techniques: "several agent-based frameworks that utilize machine learning for intelligent decision support have been recently reported" (Jedrzejowicz, 2011). In the context of learning agents, ML can be considered as a toolbox, providing various algorithms to learn behavior from data.

A vital part of agent-based approaches in manufacturing consists in the perception to determine the conditions of their environment. The usage of sensors and other devices for the gathering of data from the manufacturing shop floor delivers information that can be useful for different machine learning scenarios. Due to the perception of agents, many learning strategies pursued by agents focus on supervised learning algorithms as the sensor information provides a correct response that is used for a teaching of the agents. Thus, in terms of this work, especially in the next section, supervised learning strategies are employed to perform a predictive maintenance scenario based on sensor data from the agent's environment.

7.3 Predictive Maintenance Manufacturing Scenarios

To provide a use-case suitable for the manufacturing domain, a predictive maintenance scenario has been selected to show the advantages that are implied by enabling machine learning within the proposed agent architecture. As predictive maintenance scenarios are also of major interested in a vast number of industrial use-cases the application with learning agents provides a major benefit to state-of-the-art manufacturing systems without any prediction in terms of machine breakdowns or expected maintenance costs.

There is no general definition of predictive maintenance in the literature. However, Baidya and Ghosh (2015) provides a characterization of the term that is suitable for the application within this work: "Predictive maintenance is some form of activity aimed at identifying the presence of deterioration or defining the extent of deterioration that already exists. At the end of a predictive task, the person performing the predictive tasks knows more about the asset, but the condition of the asset has not been changed." In other words, predictive maintenance methods target to determine the health of in-service equipment, such as machines or tools, in order to predict when a breakdown is likely to occur.

The predictive maintenance concepts provides different maintenance strategies, such as time-based maintenance or Prognostic Health Management (PHM). While time-based maintenance is a rather simplistic strategy that assigns periodic maintenance intervals regardless of the assets actual health condition (Mosallam et al., 2016), the PHM represents a more sophisticated approach and refers to "a process which links degradation modeling research to predictive maintenance policies". The PHM is a vital part in predictive maintenance research and can be characterized by four main modules (Medjaher et al., 2012):

Fault detection Recognition of an error, independent of its cause. (Dong et al., 2012)

Fault diagnostics Classification of an error and identification of its cause. Choi et al. (2009)

Fault prognostics Remaining Useful Lifetime (RUL) prediction. (Tobon-Mejia et al., 2012)

Decision making Selection of an optimal time slot for maintenance. (Iyer et al., 2006)

Despite the importance regarding all of these research fields the actual prognostics plays a vital role in most manufacturing use-cases. Thus, the scenarios carried out in terms of this work concentrate on the assessment of a RUL in terms of predictive maintenance.

7.3.1 Prediction of the Remaining Useful Lifetime

The RUL prediction characterizes an estimation of the time before machine failure. In general, two main approaches can be distinguished in terms of determining the RUL – physics-based models and data-driven models. Whereas the physics-based models require an in-depth understanding of the processes and machines involved, data-driven approaches make use of sensor data to learn patterns from this data. As this work pursues an approach to enable machine learning scenarios within learning agents, the data-driven approach is selected.

In the domain of data-driven RUL, two types of predictions are mainly distinguished in the state-of-the-art literature – indirect and direct RUL prediction (see Figure 7.2). Within the direct RUL approach, the prediction is performed in two steps: (i) Determination of the machine's health status; (ii) execution of a RUL prediction algorithm, when a certain threshold of degradation is exceeded (Benkedjouh et al., 2015; Gorjian et al., 2010). The two-step approach is characterized by a determination of the health status, i.e. the degradation, based on sensor data in the first step and the derivation of the RUL in the second step. As the RUL is not derived directly from the sensor data, this approach is characterized as indirect RUL prediction.

In contrast to this indirect approach, the direct RUL prediction makes immediate use of the sensor data by deriving a RUL directly from sensor data patterns. As the direct RUL prediction offers higher potentials in terms of machine learning scenarios, e.g. to be applied in decentralized

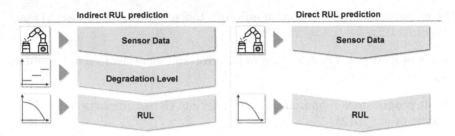

Figure 7.2: Indirect and direct RUL prediction

multi-agent system environments, this approach is selected for a further investigation in this work.

7.3.2 Direct RUL Prediction with Neural Network Approaches

The direct RUL is the most straight forward approach to determine a remaining useful lifetime for predictive maintenance scenarios directly from sensor data. The ML algorithm recognizes patterns based on the available sensor data and relate this patterns to the RUL. In order to pursue the direct RUL, different procedures can be pursued. One flexible approach that is capable of determining the RUL based on a sufficient quantity of sensor data is introduced by Heimes (2008). The proposed procedure makes use of a Recurrent Neural Network (RNN) in order to determine the best model for the RUL prediction. In order to find a good architecture for the RNN, a Genetic Algorithm (GA) is applied for the determination of the hyperparameters that are needed for the application of a ANN. By making use of this approach, Heimes (2008) was able to win a data challenge contest of the PHM Society. The methodology to determine the best architecture is shown in Figure 7.3.

On the top of Figure 7.3, the RUL prediction process is visualized. After a preprocessing of the sensor data using a *Kalman Filter* a recurrent neural network is used to determine the remaining useful lifetime. In the last step, the test score is evaluated to assess the achieved prediction quality. As the results of a RNN process are strongly depending on the hyperparameters (optimal number of layers and neurons) that are selected for the neural network model, the process needs to be executed multiple times with different parameter sets in order to deliver the desired results. Within the model described above, this is done using a genetic algorithm.

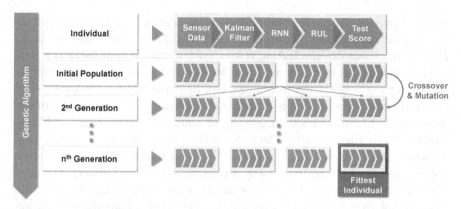

Figure 7.3: Direct RUL prediction by means of a RNN approach

The GA shown at the left side of Figure 7.3 indicates multiple life cycles of the RNN process model adapted through genetic evolution. Multiple generations of individuals representing the five-step-model are produced by crossover and mutation operations. In the nth generation the fittest individual is selected representing the best model for the RNN estimating the RUL of the predictive maintenance scenario. This proceeding using a GA also seems to be well suited for an application in dynamic environments, since the GA can be utilized to adapt the RNN to changing environmental conditions. This makes the approach especially applicable for agent-based environments, in which agents are able to sense their environment.

7.4 Application of ML Scenarios in MAS

This section focuses on the integration of scenarios such as the predictive maintenance use-case into multi-agent systems. The MAS architecture that is targeted by means of this approach is capable of autonomous agent behavior and negotiation techniques between the agents. In order to quantify the proposed scenario a modular price is introduced that enables negotiation techniques between agents of an MAS.

7.4.1 Intelligent Negotiation Techniques in MAS

The core idea of MAS in manufacturing consists in autonomously acting agents that manage the production of a good in cooperative manner. This cooperation usually manifests itself in communication and negotiation techniques that are utilized e.g. to determine manufacturing sequences or to find the most suitable production resource for a certain manufacturing step.

The mechanisms that are employed to realize a manufacturing scenarios based on these negotiation techniques can be explained with regard to the organization of a production in holonic manufacturing systems as introduced in section 4.3.1, in which the mapping of holon type functions to MAS is explained (see Figure 4.10). The order holon constitutes the interface to a customer and is accordingly represented by a customer agent. The product holon, which is either represented by a resource allocation holon or a process control holon is modeled as a coordination agent in MAS. Finally, the resource holons are represented by smart agents that pursue the negotiation techniques mentioned above. A typical configuration of an MAS that is capable of autonomous negotiation, is depicted in Figure 7.4.

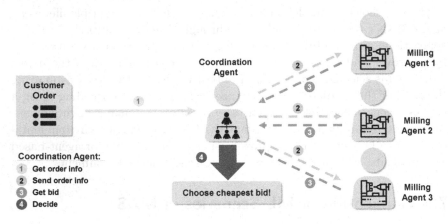

Figure 7.4: Negotiation mechanisms between smart agents in the field

The customer order is propagated into the MAS by means of a customer agent or by some other high-level system interface containing product information that are related to the according order (step 1). The coordination will receive this order further propagate the information to agents in the field (step 2). Within the example shown in Figure 7.4, the resource agents represent milling machines in the field, the smart agents are accordingly referred to a milling agents. The coordination agents receives bids from all agents that are suitable to perform the demanded manufacturing step (step 3). In this example, all agents shown will return an offer as they all represent identical manufacturing resources. The coordination agent will then evaluate the bids and will make a decision that assigns the demanded manufacturing step to the agent with the best bid (step 4).

Within this centralized approach, the coordination agent is in charge of the production steering. The according procedure is also referred to as Contract Net Protocol (CNP) that can be regarded as an auction-based market-mechanism and in which the task distribution of the process is affected by a negotiation process (Smith, 1980). In order to enable an objective decision-making whether the manufacturing step should be accomplished by a certain manufacturing resource, quantitative measures for evaluating the different bids are needed. In terms of this work, a price functions is selected in order to provide such quantitative measure.

7.4.2 Quantitative Price Function for Negotiating Agents

To assess the bids coming from different agents against each other, a quantitative price function is used that reflects different cost factors and boundary conditions which might affect the real costs of the manufacturing process. Thus, the price consolidates various KPI, e.g. such as energy costs, material costs, logistics costs or machine degradation, into one coherent number that enables a cross-agent comparison for the same manufacturing steps:

$$p_x = \frac{\sum\limits_{i=1}^{n} w_i \cdot p_i}{n} \tag{7.3}$$

where $i, \ldots, n$ represent the actual cost factors that are taken into account, p_i are the partial prices offered by the agents for each of the cost factors and w_i are weight factors that represent the importance of the according cost factor with regard the actual manufacturing process. Finally, p_x represents the bid of agent x responding to an offer from the coordination agent.

This price function can be used in order to establish a bidding process based on the CNP procedure mentioned above. The next step according to learning mechanisms of an agent is to enhance this price function to an extent that allows for an adaptive adjustment of the price that reflects not only static parameters like the weights w_i, but also additional factors that depend on the learning and the actual production history of an agent.

The modular price that is carried out for the purpose of this work takes into account such dynamic parameters by focusing on the predictive maintenance scenario in combination with machine learning algorithms that target a determination of the remaining useful lifetime of an agent. Thus, the price that is returned by an agent will also reflect whether a possible machine defects of the according production resource are likely to occur or not. Other parameters that are taken into account by the price function target the predicted quality as well as the manufacturing costs itself. The components of the price function are depicted in Figure 7.5.

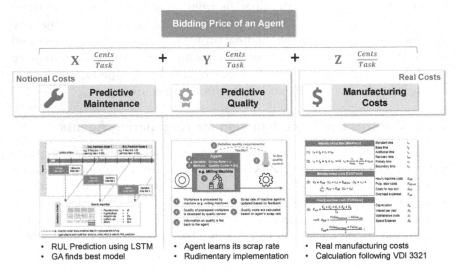

Figure 7.5: Bidding price derived from the three-module price function

The dynamic price function consists of three modules that each yield at specific costs with regard to the manufacturing process. Whereas the manufacturing costs module take into account real production costs in terms of resource consumption, human resources and machine depreciation, the predictive maintenance and predictive quality costs focus more on notional costs that take into account the risk of machine failures or breakdowns as well as risks of bad product quality. Thus, combining these partial costs with the general price function derived in equation 7.3 the equation for the full bidding price $p_{b,x}$ can be determined to

$$p_{b,x} = X + Y + Z$$

$$= \frac{\sum_{i=1}^{l} w_{PM,i} \cdot p_{PM,i}}{l} + \frac{\sum_{j=1}^{m} w_{PQ,j} \cdot p_{PQ,j}}{m} + \frac{\sum_{k=1}^{n} w_{MC,k} \cdot p_{MC,k}}{n} \quad (7.4)$$

In equation 7.4 each of the three components is constructed similar to the price function derived in equation 7.3. The predictive maintenance costs

X are determined by the according weights $w_{PM,i}$ and partial prices $p_{PM,i}$. The costs for predictive quality (PQ) and manufacturing costs (MC) are derived in a similar manner. The sum of these costs represents the actual bidding price $p_{b,x}$ of agent x.

The coordination agent only receives the bidding $p_{b,x}$ that represents all the aforementioned costs and partial costs in a condensed form. Thus, the model generation and training regarding the calculation of each price functions components belongs to the agents. This proceeding preserves the autonomy concepts and independence of each agent in the MAS.

The entire costs for the manufacturing of each product c_P are then calculated by means of the raw materials costs c_{RM} and the sum of the agents bids for each manufacturing step:

$$c_P = c_{RM} + \sum_{x=1}^{N} p_{b,x} \cdot s_{MS} \tag{7.5}$$

where s_{MS} is the number of equal manufacturing steps for the same product and N represents the total number of production resources that take part in the manufacturing of the product.

7.4.3 Predictive Maintenance Costs through Machine Learning

This section describes the actual determination of the costs induced by the risk of machine breakdown through machine learning algorithms that are able to anticipate the probability of machine failures based on historical sensor data. The sensor data that is used for this purpose aims at a prediction of the remaining useful lifetime of the according machine.

Data sets utilized for these data-driven methods can be collected from various sources, such as actual data from machines, historical data from manufacturing processes or existing data sets that are used for data challenges or other data mining related projects. For the purpose of this work, a data set from a challenge within the PHM society was selected due to a sufficient variety of run-to-failure data. The *Engine Data Set* used for this purpose covers simulations of error propagation in a turbofan engine and was investigated by Saxena et al. (2008). The actual data sets consists of 200 simulation runs by making use of the Modular Aero-Propulsion System Simulation (C-MAPPS). The sensor data is made up by means of three different engine modules and actual measurement data from 21 sensors. Thus, the data sets consists of a total number of 24 features that can be taken into account for the machine learning process. Each of these 200 simulations

contains a few hundred discrete measurement steps, in which each feature is
determined with concrete data. Due to the similarity of sensor values with
measurements from actual manufacturing machines, the data sets of the
turbofan engine are also quiet feasible for the simulation of a manufacturing
process based on a milling machine or other production resources.

The training of the RNN according to the method described in Figure 7.3
requires the remaining useful lifetime of the machine at each discrete simulation step. As the available data sets only provide sensor data as well
as the time step of the machine breakdown, an estimation of the RUL has
to be performed based on realistic assumptions regarding the degradation
process of the machine (see Figure 7.6).

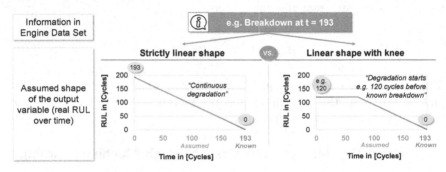

Figure 7.6: Bidding price derived from a three-module price function

The figure shows two different shapes for an estimated degradation of
the machine, both terminated by a breakdown at *cycle 193*. On the left
side of Figure 7.6, a linear/continuous degradation of the machine lifespan
is depicted, while the right diagram indicates a degradation that starts
120 cycles prior breakdown. Although both assumptions were investigated
in the data preprocessing, the assumed shape on the right side led to
better/more realistic results and was selected as model presumption for the
applied machine learning algorithms.

Data Mining and Transformation of the Sensor Data

The actual analysis of the data sets in terms of a RUL prediction is performed
by common data mining techniques according to the Knowledge Discovery in
Databases (KDD) process. The KDD process provides means for extracting
high-level knowledge from low-level information (Fayyad, 1996). The single
steps of the KDD process are shown in Figure 7.7.

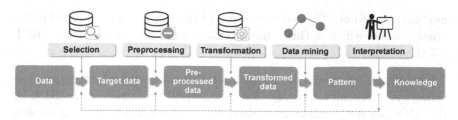

Figure 7.7: Knowledge Discovery in Databases Process as general data mining design pattern

After a selection and preprocessing of the data, the mining step is performed representing the actual application of a specific algorithm for the extraction of general patterns from the data. In terms of the predictive maintenance scenario targeted in this work, the data mining step is carried out by means of an Artificial Neural Network as described in section 7.3.2.

The data mining process shaped for the application of the RUL prediction consists in the usage of ANN in terms of a recurrent neural network evolution realized by a Long Short-Term Memory (LSTM) network. LSTM are characterized as a subtype of RNN and represent the state-of-the art for data analysis in the sort of networks (Hochreiter and Schmidhuber, 1997). In order to select suitable model parameters for the LSTM, such as the learning rate and the number of hidden units, the genetic algorithm introduced in section 7.3.2 is employed. The model evolution is finally deployed according to the process in Figure 7.3.

For the implementation of the LSTM the Deep Learning for Java (DL4J) library was utilized. Java was selected as the preferred language in order to guarantee an interoperability with other modules developed in terms of this work's framework. By making use of the DL4J framework, a LSTM network is carried out comprising of the following elements:

24 input neurons as input layer Every feature of the data set is represented by a neuron.

number of hidden units the number of hidden units that are located within the hidden layer is a variable and constitutes one of the hyperparameters of the RNN. As the hyperparamters are varied in terms of the genetic algorithm, the number of hidden units changes during the training of phase of the RNN.

one output neuron The output neuron of the RNN process represents the target value, which in this context is the remaining useful lifetime (RUL of the machine).

The number of hidden units which has to be determined during the GA process is of crucial importance for a successful prediction of the RUL by the ANN. As ANNs are characterized by an exponentially growing number of connections between the hidden units, if these units are added, risks of overfitting the modal have to be carefully taken into account. Regarding ANNs, overfitting can be thought of as a network that solely memorizes the training data and their respective labels, instead of learning the underlying patterns. Overfitting occurs if a network exhibits too many parameters relative to the number of available observations (Leinweber, 2007). If a model is overfitted, the ANN will achieve good results on the training data, but will perform inadequate on new test data. Nevertheless, if the number of hidden units in the layers of the ANN is not sufficient, the model is not able to map complex relationships and patterns within the data needed for a precise prediction of the target value.

Besides the number of hidden units in the layers of an ANN, other hyperparameters are also of major importance for the performance of the model. As the hyperparameters can be regarded as the architecture of an ANN, they need to be determined in advance to the model's training. After all hyperparameters have been specified, the model is filled with weights that represents the relations between the hidden units similar to the thickness of connections between neurons. A total number of six hyperparameters have been selected for the model generation of the predictive maintenance use-case and are defined as follows:

Number of hidden units The number of hidden units/LSTM memory blocks in the layer of the network.

Learning rate The learning rate of the training algorithm.

Seed for wieght initialization In order to train the model, random weights are assigned to each connection between the hidden units. These weights characterize the starting point for the training algorithm. This random initialization can be controlled with a seed, wherein the seed is a randomly selected number.

Initial (real) RUL The initial remaining useful lifetime in order to obtain the shape of the real RUL over time.

Number of lagged time steps The strength of the recurrency in the RNN (backwards) which is equal to the number of observations that are used for the RUL prediction.

Size of the mini batches During the training phase of the model, the weights are iteratively updated by the training algorithm. The mini batch size defines the number of observations that need to be performed before the weights are updated.

The determined model takes the sample machine data, divides these sample values into training and validation sets and accordingly determines a model that is able to predict the RUL for the indicated machine process.

7.4.4 Implementation of the RUL Prediction into Smart Agents

The implementation of the derived model and the usage of the model in terms of the OPC UA based multi-agent system is realized by an interconnection of the MAS with an analytics center. As the computational power of embedded devices in the field is not sufficient for an actual application of genetic algorithm techniques in combination with machine learning scenarios, the computational process has to be outsourced. The realization of this outsourcing strategy is carried out in the form an *Analytics Cloud* that performs the actual determination of hyperparameters through the GA as well as the model training for each individual agent (machine). The principle architecture of this solution is shown in Figure 7.8.

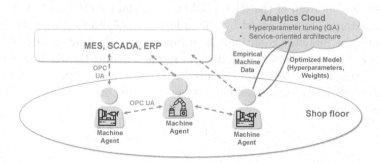

Figure 7.8: Optimization of the RUL prediction model in terms of an external cloud solution

The main idea behind this outsourcing concept is to decouple the RUL, which is required for the bidding process of the agents, from the optimization and training of the prediction model. Accordingly, each agent reserves a specific computation instance within the analytics cloud. Different threats in the cloud carry out distinctive optimization and training processes of an LSTM for each specific machine agent. Thus, the model training for each agent is carried out based on the agent's personal background, perception and historical events. The results of the RUL model optimization in the analytics cloud are specifically

- the hyperparameters defining the final shape of the LSTM for a specific machine agents,

- the weights between the neurons of the ANN.

The final model can be easily implemented into the agents in the field in terms of a simple correlation or other representation of the prediction model. After the model has been deployed on the agent's embedded devices, the agents are able to use this model in order to perform a RUL prediction before each production step. Depending on the required precision and actuality of the model, the training process of the LSTM can performed frequently, e.g. every 50 task operations of an agent and on the basis of (new) sensor data. This way, the model representation that is always up-to-date an a prediction accuracy can be guaranteed.

7.5 RUL Learning Agents in OPC UA Based MAS

The implementation of the RUL prediction model into the OPC UA based agent instances is performed by a simple OPC UA *Method* definition that contains the model correlation. As input values, the sensor values of the agents are inserted into the model. As model output, the RUL prediction values is returned. Based on this prediction, the price supplement for the implicit predictive maintenance costs can finally be added to the agents bidding price:

$$p_{PM} = p_{max} \cdot \left(1 - w_{\text{RUL}} \cdot \frac{\text{RUL}_{pred}}{\text{RUL}_{init}} \right) \tag{7.6}$$

where p_{PM} is the price derived from the predictive maintenance costs. p_{max} represents the maximum penalty price in the event of an actual machine breakdown, while w_{RUL} is a weight parameter that indicates the influence of the RUL on predictive maintenance costs.

8 Use-cases for Industrial Automation Processes

The use-case section of this work demonstrates the potentials of the derived framework in terms of real-world applications that make use of the scalability concepts of OPC UA based MAS. One major aim of these applications is to show the extensibility of the derived framework in terms of domain-specific information models. The applicability to solve practical problems that are of interest in production and manufacturing organization is shown in terms of two use-cases, of whom one resulted in the development of an *Industry 4.0* testbed that was demonstrated at the Hannover Fair 2016.

Section 8.1 focuses on the application of the derived multi-agent system concepts on an existing *Industry 4.0* scenario that has been carried out in a joint project among multiple universities across Germany and which is known as the "myJoghurt" demonstrator.

Section 8.2 demonstrates the extensibility of the object-oriented modeling concepts for a manufacturing related use-case that uses rich information modeling capabilities offered by the domain-specific model extension using OPC UA. To show the applicability of the domain model in terms of an autonomous production planning and execution within a manufacturing environment, a demonstrator has been carried out that shows the potentials of OPC UA based MAS in terms of the manufacturing of a customized products toward a "Lot-size 1" targeted production.

Section 8.3 demonstrates an *Industry 4.0* testbed that using OPC UA based MAS in connection with a software demonstrator, in which distributed embedded systems emulate the manufacturing of a customized product based on sole communication with each other. In this use-case, information exchange with high-level planning and customer-related HMI is carried out by extension of the ICT infrastructure using the MQTT protocol.

© Springer Fachmedien Wiesbaden GmbH, part of Springer Nature 2019
M. Hoffmann, *Smart Agents for the Industry 4.0*,
https://doi.org/10.1007/978-3-658-27742-0_8

8.1 Domain Ontology for the "myJoghurt" Testbed

Due to an increasing interest in *Industry 4.0* scenarios that demonstrate the potentials of digitized factories the elaboration of suitable testbeds/demonstration scenario is an essential part of today's research on manufacturing. One of these demonstrators has been carried out by several German universities, of whom the most remarkable are the Technische Unversität München (TUM), the universities of Augsburg and Stuttgart as well as the RWTH Aachen University. The testbed scenario is referred to as "myJoghurt" and was originally carried out by the Institute of Automation and Information Systems (AIS) at the TUM. One major feature of the testbed is that its physical and software components are distributed among different locations all over Germany. Thus, different parts of the manufacturing and organization processes are coordinated using a decentralized approach (Mayer et al., 2013). The aim of the testbed is to show the potentials of an agent-based based reconfiguration of distributed, interconnected, smart production units. The demonstrator shows an information-technological coupling of spatially separated factory units that are able to adapt autonomously with regard to the number of production facilities and to dynamically scale the production using additional manufacturing resources (Pantförder et al., 2014).

Further target objectives of the demonstrator are to show the potentials of autonomous software agents that are able to organize the production process independently and without human intervention. The underlying MAS organizes the production in an autonomous way while dealing with errors or possible malfunctions in an intelligent manner. The considered production flow is mapped from the start the start of the production, which is initiated by accepting orders of possible customers, to the delivery of a final product (Mayer et al., 2013).

The contribution of this work with regard to the demonstrator consists in an enhancement of the communication capabilities between the smart software agents by establishing an OPC UA based MAS. Thus, after a brief introduction of the demonstration process and the according work flow, an OPC UA information model is derived in form of a companion specification that maps the "yogurt" domain targeted by the demonstrator.

8.1.1 myJoghurt Demonstration Scenario and Work Flow

The "myJoghurt" demonstrator represents a manufacturing line producing customized yogurt. It consists of four asynchronously executed manufacturing steps, which involve the *yogurt production, yogurt refinement*, a *cap engraving facility* and the *filling* of the yogurt (see Figure 8.1).

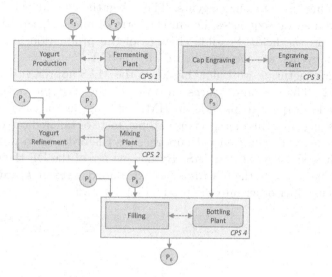

Figure 8.1: Process description of the myJoghurt production line (Weyrich et al., 2014)

The first step consists in the yogurt production within a fermenting plant. The raw materials of this step are milk (P_1) and bacteria called yogurt culture (P_2). The semi-finished product (P_7) is transported to the yogurt refinement plant, in which additional ingredients like toppings (P_3) are added. Parallel to the refinement in the mixing plant, the engraving of the yogurt's cap (P_5) is performed by means of an engraving plant. In order to perform the final production step – the bottling of the yogurt – three components, the bottle (P_4), the finished yogurt (P_8) as well as the engraved cap (P_9) are needed. Thus, in the bottling plant, these components are finally *assembled* to form the finalized product (P_6). Each of the described production steps can be carried out by various redundant process cells depending on their availability. Each of the process cells is characterized as a CPS that consists of an intelligent, embedded device (agent) attached to the according machine. Every agent registers itself to the AMS and informs the DF about its services or capabilities (Weyrich et al., 2014).

The production process in the presented system architecture is carried out by means of a *coordination agent (CA)* that manages the process and performs a selection of suitable agents. The CA represents the central agent of the MAS and is the single instance accepting customer orders. After a reception of these orders the CA has the capabilities of planning, scheduling and assigning tasks to single agents. The CA gathers all knowledge about yogurt production sequences, i.e. in the form of an ERP, and supervises all agents in the field. Possible changes of the production, e.g. due to unforeseen events, are also coordinated by the CA.

The manufacturing work flow based on the MAS approach is shown in Figure 8.2. The customer places an order to the *customer agent* which has the sole purpose to provide an HMI, offering design capabilities for customers, e.g. in terms of a specialized topic, taste or tailored cap engraving (`Place Order`). The customer agents receives the order via a web service, however it is also part of the MAS. By making use of the DF the customer agent is able to locate the CA (`Get/Return Coordination Agent`) and to pass the customer order into the production network.

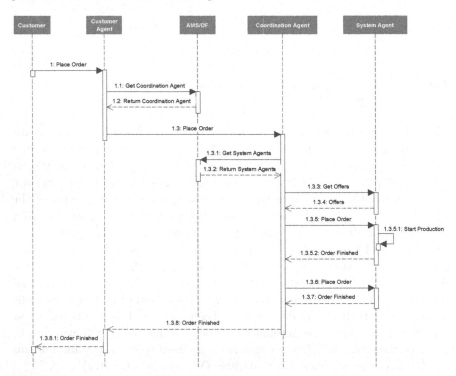

Figure 8.2: Sequence diagram of the ordering and production process

The coordination agent analyzes the order request and extracts the information that is relevant to the production process. In the next step, the CA splits the according production plan into single tasks that can be accomplished by one the available production cells (**Get/Return System Agents**). In order to plan and schedule the production, the CA needs to select suitable agents. This is performed by means of a bidding mechanism similar to the one explained in the machine learning chapter. In terms of the myJoghurt process a simplified mechanism is used, in which the CA sends offer requests (**Get Offers**) and the each of the agents returns an according price for the production steps they are able to perform (**Offers**).

After the CA has received the offers of all agents that are willing to perform at least one of the required tasks, the CA places manufacturing orders to the single agent instances (**Place Order**). The agents in the field accordingly perform the assigned tasks (**Start Production**) and report the finished production steps to the CA (**Order finished**). The coordination stores this information and checks, whether the next production step in the manufacturing sequence can be initiated. While the yogurt refinement step depends on the finishing of the yogurt production, the cap engraving step can be performed independently of all other preliminary steps. The bottling step of the yogurt requires that all previous tasks have been finished successfully. After all orders are done, the CA sends a report to the customer agent, which is then informs the customer about the completion of his order (**Order finished**).

The communication in the myJoghurt MAS is carried out by a message protocol that is compliant to the FIPA ACL standard and at the same time offers interoperability to low-level controllers. The Joghurt-Produktions-Protokoll (JPP) constitutes a flexible protocol solution that operates among the three top layers of the TCP stack (*Layer 1 – Base Layer, Layer 2 – Service Layer, Layer 3 – Data Layer*). A typical message is shown in Listing 8.1.

Listing 8.1: Example of a JPP message

```
1  JPP2
2  78
3  --
4  Agent13
5  GETOFFER
6  ;
```

The first line indicates the protocol version, which is currently "2.0". The second line shows a sequential number used as an identifier to associate requests and responses. The third line contains the length of binary data in the payload. The hyphen indicates that the message content is characterized as plain text. The fourth line contains the sender, whereas the fifth line contains the actual payload of the message, which in this case, indicates that an offer request is placed to the receiving agent. (Mayer, 2014)

Although the JPP protocol solution is light-weight and therefore quite applicable in terms of low-level devices, the provided communication solutions also poses certain drawbacks. Firstly, the protocol does not support features such as authentication or encryption. Secondly, the MAS does not act entirely autonomously due to the coordination agent that organizes the production and pursues global production plans in terms of centralized system. Finally, the protocol is precisely tailored to the application domain of the yogurt manufacturing. By mapping the communication of the according software agent in terms of an OPC UA based MAS, these drawbacks can be solved by means of the scalable and secure OPC UA standard in combination with domain-specific information modeling.

8.1.2 OPC UA Information Model for the Mapping of Agent Communication

In order to fulfill the production demands of the myJoghurt use-case by making use of OPC UA based MAS the according domain ontology as well as the communication capabilities of the existing agents have to be mapped in terms of an OPC UA information model. The according information model is carried out by means of extending the OPC UA base information model as well as the companion standards for multi-agent systems and agent-based communication derived in chapter 5. Additionally, the domain-specific specification will be used to extend the existing information models in terms of knowledge about the yogurt domain in accordance with the requirements and semantics defined by the JPP protocol standard.

For the realization of the use-case by means of an OPC UA based MAS, the central messaging system described in section 5.3.4 is employed. The implementation of the messaging mechanisms is carried out through a central OPC UA server that is used for a mediation of messages between agents. This server incorporates functionalities of the white/yellow page services and at the same time can be designed redundantly to minimize the risks of a single point of failure. All agents are equipped with OPC UA clients to make use of the messaging mechanism. The realization of this message exchange principle is shown in Figure 8.3.

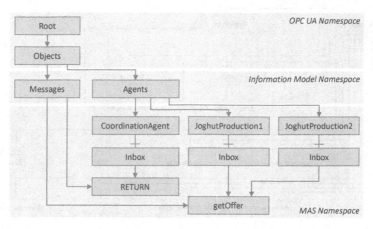

Figure 8.3: Messaging mechanisms in the yogurt application domain

For each message that is send throughout the OPC UA based MAS a message object is created in the *AddressSpace* of the OPC UA server. The agents, which are also located in the *AddressSpace*, represent different agent types containing separate inbox folders for each agent. For each message that is created during the communication process the according references from the messages folder as well as from the addressed agent's inbox folders are created. Within the example shown in Figure 8.3, a `getOffer` message is propagated to multiple agents in the MAS and a `RETURN` message is sent back to the coordination agent. The references point from the receiver to the according message object. This way, arbitrary agents are able to share information through OPC UA based messages.

Yogurt Domain Model – MessageTypes

In order to map the semantics of messages exchanged in terms of the "yogurt" domain the according *MessageType* definitions is carried out. Figure 8.4 summarizes the available message types including important properties and their integration into the *AddressSpace*.

At the bottom of the information model representation the *MessageType* subtype definitions are located representing the according message instance blueprints that are required for the communication logic in the MAS. Two additional properties are added to the general *MessageType* definition,

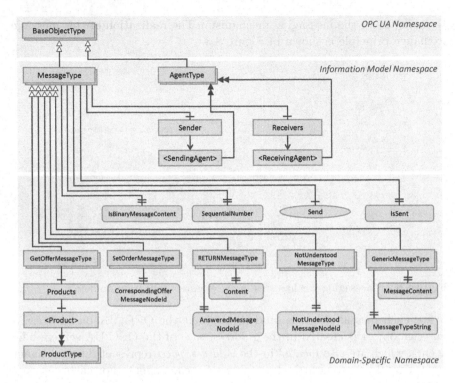

Figure 8.4: OPC UA *MessageType* definitiosn for a semantic integration of the yogurt domain

namely the `IsBinaryMessageContent` and the `SequentialNumber` property. The first one represents the JPP message part, in which the payload of a message can be declared as binary, the latter represents a consecutive sequence number the messages.

The subtype definitions are characterized by further subtypes that incorporate information models for the mapping of the agent's logic modules. Due to the complexity of these modules not all *MessageType* subtypes can be examined at this point. Exemplarily, the *GetOfferMessageType* is described here containing detailed information about the product.

The GetOfferMessageType Definition

The *GetOfferMessageType* is an integral part of the multi-agent system's logic as any *getOffer* message instance contains a detailed description about the product that is sent to the according agents in the field in order to request their capabilities. A `getOffer` request in terms of the JPP format from the coordination agent to the field agents is shown in Listing 8.2.

Listing 8.2: The `getOffer` message for the production of a customized yogurt

```
 1  JPP2
 2  24
 3  -
 4  CustomerAgent
 5  getOffer
 6  <produkt name="Joghurt">
 7      <eigenschaften kategorie="Joghurtherstellung">
 8          <merkmal id="1" name="Rahmstufe" value="___"/>
 9          <merkmal id="2" name="Laktosefrei" value="___"/>
10          <merkmal id="3" name="Milchsorte" value="___"/>
11          <merkmal id="4" name="Qualitaet" value="___"/>
12      </eigenschaften>
13      <eigenschaften kategorie="Joghurtveredelung">
14          <merkmal id="5" name="Obstsorte1" value="___"/>
15          <merkmal id="6" name="Obstsorte2" value="___"/>
16          <merkmal id="7" name="Mischung" value="___"/>
17          <merkmal id="8" name="Topping_Muesli" value="___"/>
18          <merkmal id="9" name="Topping_choc" value="___"/>
19          <merkmal id="10" name="Topping_Streusel" value="___
            "/>
20          <merkmal id="11" name="Topping_Nuesse" value="___"/>
21      </eigenschaften>
22      <eigenschaften kategorie="Deckelgravierung">
23          <merkmal id="12" name="Deckelart" value="___"/>
24          <merkmal id="13" name="Text" value="___"/>
25          <merkmal id="14" name="Schriftart" value="___"/>
26          <merkmal id="15" name="Schriftgroesse" value="___"/>
27      </eigenschaften>
28      <eigenschaften kategorie="Abfuellung">
29          <merkmal id="16" name="Behaelterart" value="___"/>
30          <merkmal id="17" name="Etikett" value="___"/>
31          <merkmal id="18" name="Verpackung" value="___"/>
32          <merkmal id="19" name="Versandart" value="___"/>
33      </eigenschaften>
34  </produkt>;
```

The `GetOfferMessage` instance constitutes a service request that is sent to each agent and demands for an offer from suitable agents in form of a price. Practically, each agent that receives the according message, is able to extract the capability requirements that are needed to fulfill the required production step and can decide whether it wants to carry out the requested task. The coordination agent that sends out the `getOffer` request expects a RETURN message in the form of a price.

The according OPC UA mapping for the content of the XML message has to represent the semantics of the message's payload in terms of the yogurt domain, thus the different categories and properties of the yogurt need to map those of the OPC UA domain-specific information model. The *GetOfferMessageType* in accordance with the OPC UA agent model and the related MAS information models is shown in Figure 8.5.

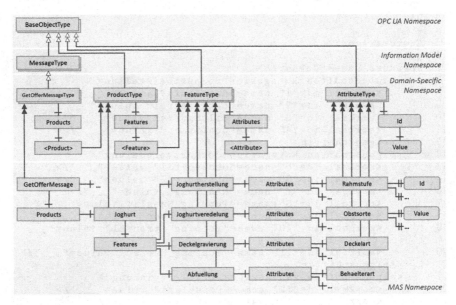

Figure 8.5: *GetOfferMessageType* specification to exchange production-related information

The *GetOfferMessage* instance contains a list of products, e.g. a yogurt. This *Joghurt* object is further characterized by a number of *Features*. Each of these features depends on a certain production cell, i.e. providing manufacturing capabilities such as yogurt production (*Joghurtherstellung*) or yogurt refinement (*Joghurtveredelung*). Each of these manufacturing steps is further characterized by a list of *Attributes* that are customized during the order process to create a unique product (*"lot-size 1"*). These attributes represent the properties of the product or its components. In terms of the yogurt, this might be the taste of the yogurt, the particular type of fruit (*Obstsorte1*) or the container type (*Behaelterart*). In accordance to the OPC UA modeling principles, each attribute is characterized by a distinctive Id and a certain Value. Each of the components being part of the *GetOfferMessage* inherit their semantics from their according type definition, i.e. *ProductType*, *FeatureType* and *AttributeType*.

The AgentType and Domain-Specific Subtypes

Within conventional multi-agent system architectures the information from and about the agents is stored among various entities. For example, the address of an agent or its current state is stored by means of the AMS, agent capabilities are managed by the DF and messages are only readable by the receiving agent. In terms of the OPC UA based MAS approach, every piece of information is globally available through the OPC UA *AddressSpace*, which can be located in a central server or on several instances according to redundancy concepts for further robustness. Each agent – by means of its representation and its knowledge – is stored in the *AddressSpace* in terms of *Agent* object. Each of these objects represents an instance of specific *AgentType* definitions. The different *AgentType* subtypes and their representation in the OPC UA *AddressSpace* for the "yogurt" domain is shown in Figure 8.6.

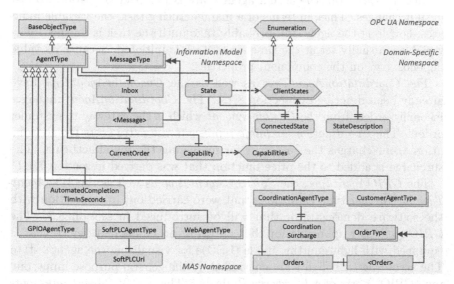

Figure 8.6: *AgentType* subtypes and their allocation in the "yogurt" model
        AddressSpace

Each *Agent* object that is instantiated from one the agent type definitions is characterized by a *State Variable* and an *Inbox* folder. The list of messages that are located in the *Inbox* are all derived from the *MessageType* definition, the possible values for the *State Variable* are derived from en `Enumeration` that contains all possible *ClientStates*. In the current configuration on the OPC UA based MAS for manufacturing systems possible agent states are *Stopped, Connected, ConnectionLost, Error* and *Working*. Additional properties of the complex *State Variable* are e.g. the *ConnectedState* and the *StateDescription Property*.

Every agent further contains a *Property* that stores the *CurrentOrder*, which is an `ID` that characterizes the manufacturing step the agent is currently performing. If the *Value* of an agent is set to *Working*, the *Value* if the *CurrentOrder Property* contains the *NodeId* of the `SetOrder` message object for the task assignment. The *CurrentOrder Property* of the coordination agent always points to the *GetOfferMessage*.

The *Capability* of an agent is selected from an `Enumeration` that contains all available capabilities. These abilities can be extended by domain-specific information models, e.g. in order to represent specialized capabilities of certain domain agents. The *AutomatedCompletionTimeInSeconds* charac-terizes the duration, the actual agent state is prospectively going to last until it changes. Thus, in terms of a manufacturing task, the variable indic-ates, how long the agent will probably take until the task is finished. The variable is usually set at the time a new task is initiated, because its value depends, e.g. on the component properties.

The *CoordinationAgentType* as well as the *CusomterAgentType* have already been described in section 8.1.1. The *CoordinationAgent* manages incoming orders from the *CustomerAgent* which is the gateway to customer orders. The *CoordinationSurcharge* represents the extra costs the coordin-ation agent charges the customer for managing the manufacturing. The surcharge is added to the price function that was derived in section 7.4.2.

The *GPIOAgentType*, the *SoftPLCAgentType* as well as the *WebAgent-Type* are further type definitions that were carried out in connection with the software demonstrator that will be introduced in section 8.3. The *GPIOAgent* is responsible for the control of an embedded device that is equipped with light-emitting diode (LED)s for visualizing the agent's state. The according LED lights are controlled by the general purpose input/out-put (GPIO) ports of a *Raspberry Pi* device. The *SoftPLCAgent* represents a simulated PLC running a *CODESYS* software stack (CODESYS, 2017). Finally, the *WebAgentType* represents an agent that is responsible for the interconnection of the MAS with a web-service enabled web application.

The web connectivity module of this agent is will be further explained in section 8.3.

8.1.3 Achievements and Limitations of the Use-Case Design Model

The yogurt domain model shows the information modeling capabilities of OPC UA in connection with the basic companion specifications for the mapping of MAS that were carried out in chapter 5. The use-case clearly shows that the mapping of manufacturing resources to production steps of a customized product can be carried out with a high flexibility and adaptability. Further, the OPC UA based design approach is capable of mapping domain-specific information, e.g. with regard to a certain product family, on generic agent types representing arbitrary resources in the field. The modeling methodology of the specialized agents, orders and product types shows that the approach allows for generic scalability of the models.

Limitations of the use-case might consist in the autonomy of the MAS that was tailored to the yogurt domain. The coordination agent is characterized as a single point of knowledge, whereas the autonomy of the agents in the field is limited to an indirect negotiation mechanism in terms of a price bidding. Although the approach described already characterizes a powerful MAS, the use-case was further developed and transferred to a manufacturing related use-case. This use-case takes into account the MAS approach demonstrated in this chapter, but also takes into consideration the intelligence and learning approaches from chapter 7 and the connectivity modules to high-level systems from chapter 6.

8.2 Manufacturing Use-Case

This section focuses on a use-case that is directly related to the manufacturing domain. The design model of the domain-specific model is derived similarly to the yogurt domain, the manufacturing execution however, is carried out by means a decentralized control scenario rather than through a central agent that coordinates all processes.

8.2.1 Process Description

The demonstration scenario models a simplified manufacturing process that is characterized by two components that are simultaneously produced before an assembly process composes the final product (see Figure 8.7).

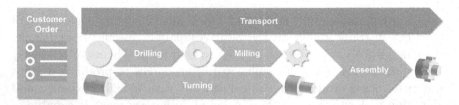

Figure 8.7: Demonstration scenario of the manufacturing use-case

The customer order is carried out similar to the customized yogurt. After specifying the production steps of the manufacturing line, the raw materials of a *gear* and a *shaft* are automatically transported the according machines that begin with the production. The *drilling* step drills a hole into the raw material of the *gear*, whereas a *turning* machine processes the raw material of the *shaft* into the desired form. After the the gear is processed by the *milling* machine, the finished *gearwheel* as well as the shaft are transported to an *assembly* robot that assembles the product into its final state.

8.2.2 Autonomous Organization of the Production Process

The organization process of this manufacturing line is organized based on a cooperation of the agents with each other. The knowledge of the agents consists of the perception of their environment, i.e. sensors attached to the machines they represent, and additional information from the gateway agent to an ERP system. The process knowledge that is present in the ERP system becomes accessible to each agent according to the blackboard pattern, whereas the OPC UA *AddressSpace* represents the blackboard. In order to enable this way of retrieving information from the ERP a set-up of the ERP is initiated during start-up of the MAS.

Figure 8.8 shows the initialization of the OPC UA gateway agent and the *AddressSpace* adaptation during start-up of the ERP system. The Ofbiz system represents the open-source ERP that was used in terms of the demonstration process, the Opcclient represents the gateway agent to the ERP and the ERP-ASB constitutes the AddressSpaceBuilder module that is used for a creation of the required nodes needed for a mapping of

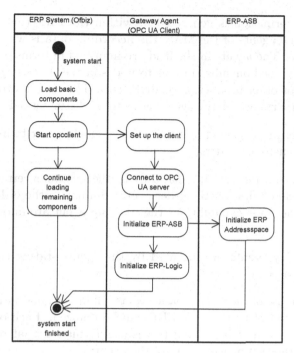

Figure 8.8: Start-up process of the ERP system and OPC UA *AddressSpace* modification

the ERP system functions. The initialization of the OPC UA *AddressSpace* is triggered by the gateway agent during immediately after the agent has connected to the OPC UA server.

Negotiation and Production Planning

After the *AddressSpace* including gateway agents and required ERP function mappings is set-up, the MAS is able to receive orders. The receiving of orders is realized through automatic events, the negotiation process between the agents that compete for each production step is carried out in terms of the price function that is dynamically adapted according to the individual learning of each agent.

The execution of scheduling and task operations with regard to the individual production steps that are needed to fulfill the manufacturing of the product is autonomously organized by the agents. In order to realize this autonomous behavior, an event-driven approach is utilized. Thus, at any time, information is generated in within the ERP that could be of interest to the agents of the MAS, the according data is written into the *AddressSpace*. The agents in the field are informed in terms of events that are triggered based on subscriptions to nodes in the *AddressSpace*, such as task folders or other production-related items. Three main events that are of special importance for the agents have to be distinguished:

Order events are triggered by the ERP systems in order to inform agents about new customer orders.

Task events are triggered automatically by the agent of a certain capability that has reacted first on the order event. Similar to the order event, all agents of the same capability as the reacting agent are informed by the according task events.

Issue inventory events are initiated by the agents stating that new raw material is needed.

By making use of these three basic events, all major information exchange scenarios of the MAS with the ERP can be realized. Each of the events hereby contains three important pieces of information that represent the context information or meta data of the event:

Event type The event type specifies the purpose of the event, simple classifying the event whether it can be characterized as *order*, a *task* or an *inventory issue*.

Origin The origin of the event characterizes the source of the triggered action. This information is especially important for an identification of the correct items in the *AddressSpace*.

Event cause The third additional piece of information is wrapped into a short message that informs all receiving agents about the reason the event was issued. By making use of this information the agents can anticipate whether it is necessary to browse the *AddressSpace*. This behavior is especially important to optimize the MAS *performance*.

Due to the absence of a central coordination agent, the agents need to react autonomously on order events coming from the ERP. Due to a simultaneous triggering of the according order events at all available agents, concurrency problems between agents with the same capabilities trying to process the order at the same time, are likely to occur. In order to bypass this issue, a ranking mechanism is carried out at each start-up of the MAS. The ranking mechanism assigns a consecutive rank number to each agent of the same capability in the range from $0 \ldots n$, where n represents the number of the nth agent with the same capability.

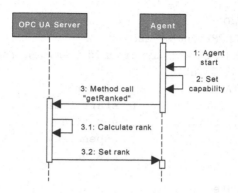

Figure 8.9: Demonstration scenario of the manufacturing use-case

Figure 8.9 shows the continuous assignment of rank numbers to agent immediately after start-up and allocation of the according capability to the agent. Based on this ranking, the processing sequence of incoming order events is clearly defined. In the first attempt, the agent with the lowest rank number – likely agent 0 – strives for a processing of the according event. Only in the case that the agent is disconnected, occupied are not available due to some other reasons, the processing will be continued from the agent with the next rank number $i + 1$. This way, concurrent processing of the same order is prevented under all circumstances.

After processing of the order event by all capability agents with the lowest ranks, the according agents split the order into items and into tasks that represent all capabilities and actions that need to be performed for the production of each item. The representation of the tasks within the OPC UA *AddressSpace* is shown in Figure 8.10.

It is of major importance that the processing of orders and their split into distinctive tasks does not decimate the semantic context of the order or erases important meta information that is related to the product. Due to this requirements, the tasks that are stored into the *AddressSpace* (left side of Figure 8.10) are created in terms of a consistent naming schema that preserves all information that is needed for the product to be manufactured (right side of Figure 8.10). Accordingly, the name of each tasks is composed

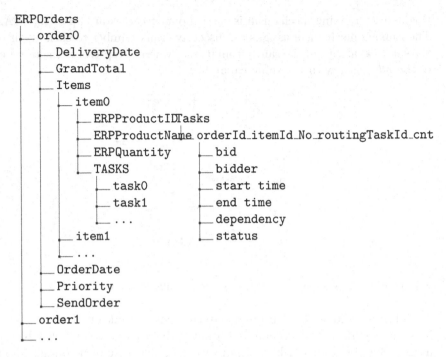

Figure 8.10: Event-based orders and task representation in the OPC UA *AddressSpace*

of the *OrderId*, *ItemId*, the sequential number of the item *No*, the *RoutingTaskId* and the *cnt* variable which represents the task counter. By means of these descriptive pieces of information it is always possible to precisely relate a manufacturing task to an existing order and the according product.

Bidding Process and Task Execution

The bidding mechanism is carried out by an autonomous price adaptation of the agents. After receiving the order as shown in Figure 6.4, processing of its content and completed split into distinctive tasks, the agents start the bidding process (see Figure 8.11).

The bidding process is initiated by the finishing of a task from a previous agent. The determination of *previous* production steps is performed using the product routings that are located in the production description of the *AddressSpace* (see Figure 6.7). In the next step of the bidding process,

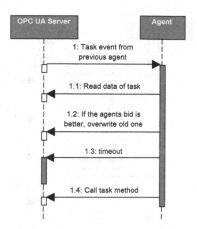

Figure 8.11: Task bidding process

the agent reads information about the task. This information includes data about the capabilities that are needed to fulfill the process step and additional information about the current bidding price (see Figure 8.10 right side). If the bid of the agent is lower than the currently available price (bid), the agent will overwrite the existing bid and place its entity name (bidder) into the *AddressSpace*. After a certain time period – the timeout – each agent that placed a bid on the according task checks the actual name of the bidder in the *AddressSpace* object of the task. The agent with the lowest bid will find its name in the task description and will accordingly start the task execution.

The task execution is shown in Figure 8.12. After the agent has read the task information, the production process is initiated and the working status in the *AddressSpace* is changed to workInProgress. The agent sets its internal state to working and executes the task until its finished. After the process is done, the task state is changed to completed and the agent state is set back to connected. The agent is ready for the next task.

The benefits of this integrated planning, decision-making and manufacturing execution are examined in the next section that covers the development of a software demonstrator that is equipped with distributed intelligent devices that autonomously organize the production.

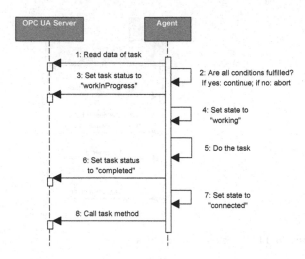

Figure 8.12: Task execution process

8.3 Demonstration Scenario – The Industry 4.0 Testbed

One major strength of derived multi-agent system concept based on OPC UA is its scalability and adaptability, especially in distributed systems. The intelligent agents – each making decisions on its own – cooperate with each other and achieve a form of higher intelligence. The application of this concept to a physical demonstrator is shown in this section.

8.3.1 Technical Setup and Realization of the Demonstrator

The architecture of the demonstrator is carried out in terms of a distributed network. Each agent as well as a central OPC UA server instance are located on separate physical machines. Various additional interfaces and system APIs are added to the architecture in order to provide an external access from a customer to the agent-based system (see Figure 8.13)

The agents shown on the left are located on the field level representing manufacturing machines. Within the demonstrator environment, the actual manufacturing process is simulated through light-weight computer devices (*Raspberry Pi*). Each of these devices is equipped with a Java-based OPC UA framework that enables OPC UA client as well as OPC UA server instances on the hardware. In terms of the functional setup of the MAS each agent

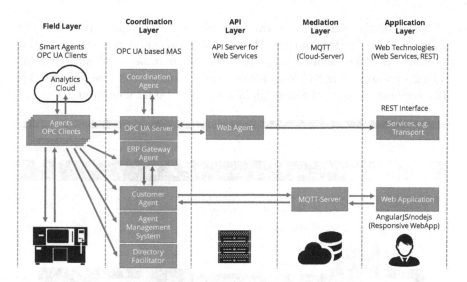

Figure 8.13: Architecture of the physical multi-agent system demonstrator

can be sufficiently represented in terms of an OPC UA client, because the central messaging server system (see alternative 3 in Figure 5.5) is used. The communication and interaction capabilities of the agents are realized through an OPC UA server that is located in the *Coordination Layer* of the architecture. The manufacturing process can be likewise coordinated through a centralized approach (production planning and scheduling by the *Coordination Agent*) or through a decentralized approach (autonomous planning by using the *ERP Gatway Agent*).

The customer agent is responsible for the acceptance of orders. The information exchange between this agent and the customer is realized by means of an MQTT server that reserves a communication channel for orders. Additional channels are carried out to provide an API to certain KPI numbers for proprietary systems. In terms of the demonstrator, all externally accessing systems (order draft and visualization of KPI) are realized by means of web applications. The web application developed specifically for the demonstrator is designed using responsive design patterns to fit on all kind of devices (computer browsers, tablets, smartphones, etc.). In terms of the software stack for these web applications, a *AngularJS / Backbone & Marionette* framework was used for the frontend, *NodeJS* was utilized for the backend.

To initiate the order process, the customer is able to tailor a product to its design wishes. For this purpose the web application offers a customization interface, in which the user is able to adjust the parameters of each manufacturing step according to his preferred product properties (see Figure 8.14). The screenshot shows the web application on a tablet computer.

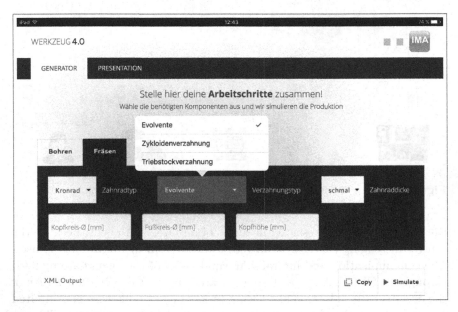

Figure 8.14: Web application for customizing the product

For each step of the production (*Drilling* – German: *Bohren*), (*Milling* – German: *Fräsen*), (*Turning* – German: *Drehen*) and (*Assembly* – German: *Montage*) the user is able to customize multiple parameters of the manufacturing process. For example, on the web page that is shown, the customer is able to change the parameters of the *Milling* step by choosing the gear type (*Zahnradtyp*) or the shape of the gear teeth (*Verzahnungstyp*). Additionally, the user might customize the inner and outer diamgeter (*Fußkreis* and *Kopfkreis*) of the gear. All customization opportunities for the manufacturing use-case are summarized in Figure 8.15. The generated product description is submitted as an XML file according to the schema in Listing 8.2 and is transferred into OPC UA objects by the customer agent.

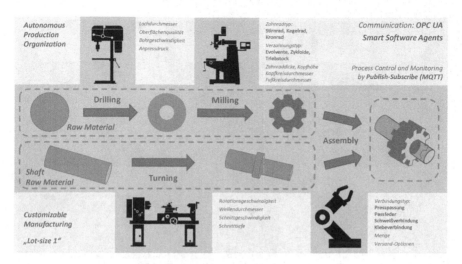

Figure 8.15: Customization of parameters for the manufacturing use-case

8.3.2 Simulation of the Manufacturing Process

After the user has customized the product according to his wishes, he is able to submit the order to the demonstrator and initiate the simulation. While the simulation is running the user is able to follow the progress of the production through the web application (Figure 8.16).

The web application screenshot shows a snapshot of the simulation process during operation. The *Drilling (Bohren)*, *Turning (Drehen)* and the *Milling (Fraesen)* machines have already finished their production steps. The transport from the *Milling (Fraesen)* machine to the *Assembly (Montage)* has just been completed for a virtual price of 0.21 $, the *Assembly (Montage)* step is in working state. The duration of the ongoing production step is also visible at the bottom right of the *Assembly (Montage)* box, currently set to 27 seconds.

The actual manufacturing simulation, which is located on the embedded computers in the field / on the OPC UA server, is constantly synchronized with the web application through the MQTT server shown in Figure 8.13. Through the publish-subscribe mechanism of the light-weight MQTT protocol, near real-time reconciliation can be achieved. As seen on the web page, the application shows all manufacturing steps as well as the actual manufacturing parameters that were chosen by the customer. The view also contain additional information such as a *Timeout* that indicates the pro-

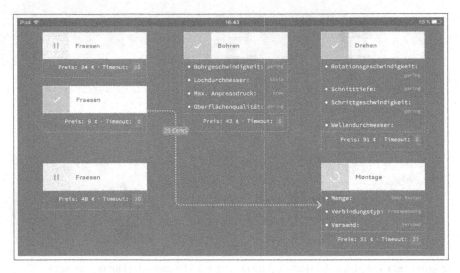

Figure 8.16: Real-time production tracking using the web application

spective remaining duration of the actual production step as well as bidding costs for each manufacturing step and transport operations.

The physical demonstrator is further equipped with additional hardware features that consist in machine models that were designed and shaped by making use of 3d printing technologies. Furthermore, the demonstrator contains two conveyor belts that literally move transport vehicle models to simulate transport processes from one machine to another. A physical control panel attached to the hardware demonstrator contains LEDs for a visualization of the agent's states (*green – ready, blue – working/occupied, red – error state*) and buttons for enabling interaction of the user with the demonstrator. For further details on the demonstrator and its physical set-up, the interested user might refer to Hoffmann et al. (2016c) and Hoffmann et al. (2016b).

As previously mentioned, the demonstrator additionally contains a physical control panel with buttons for user interaction. Each of the six manufacturing machines that are simulated throughout the process provide to interaction capabilities, a red button and a green button. The red button allows for a simulation of a machine failure or breakdown. The green button for each machine enable the user to end ongoing processes on a certain machine immediately in order to simulate an unpredicted duration for a single process step. By making use of these interactions, the demonstration use-case illustrates features that underline the applicability of the demonstration scenario in terms of an *Industry 4.0* use-case:

- The user presses the green button at one of the machines that is currently in *working* state, thus simulating a manufacturing action:

  1. The machine agent realizes the premature finishing of the production step and changes its state to *ready*,

  2. the according agent writes its new status to the *AddressSpace*,

  3. other agents, which are responsible for the transport processes as well as for further production steps of the same product react by initiating actions and start with a negotiation for the next production step,

  4. the production is dynamically reconfigured and will be continued with a time shift towards an earlier finishing of the product.

The second user interaction scenario that is realized by means of the demonstrator leads to a more complex resolution of a new situation in the manufacturing work flow:

- The user presses the red button at one of the milling machines that is currently in *working* state:

  1. The machine agent changes its status to the *error* state and stops simulating the production step,

  2. the agent writes its new status to the *AddressSpace* of the OPC UA server,

  3. using OPC UA based subscriptions to the ERP *Namespace* of the *AddressSpace*, the other agents realize that an unfinished manufacturing step has been canceled,

  4. agents with the same capability as the broken machine start the bidding process intending to compensate the breakdown of the milling machine,

  5. after the negotiation process between suitable agents, the best bid is selected to continue the canceled production,

  6. a dynamic rescheduling of the product routing is realized by the ERP system,

  7. a transport from the broken machine to the compensating agent is autonomously organized and initiated,

  8. the production continues with a new routing of the product.

These production organization and reorganization scenarios are carried out autonomously by the agents through negotiation and cooperation with each other and by means of the interconnection to an open source ERP system that provides *global* knowledge of the production. All information, i.e. simulated sensor data from the field as well as messages from other agents and additional information from the ERP is available in a well-structured form and provided by the *AddressSpace* of an OPC UA server. This way, the incorporation of field level data and high-level planning data has been successfully implemented an can be used for the production and flexible execution of real manufacturing use-cases.

8.3.3 Learning and Evolution of the Software Agents

The learning and evolution of the field agents is realized by means of the machine learning applications that had been derived in chapter 7. The bidding of the agent is specifically influenced by the price functions given in Equation 7.4 and 7.5.

As the agents within the demonstrator environment are not pursuing a real manufacturing process, sensor data is simulated in the form of suitable data samples and simulation servers. This sensor data is accordingly used within the agent's perception as low-level production data. The machine learning model for the optimization of the price function that takes into account predictive maintenance costs is updated frequently using new data. The training process of the LSTM is carried out by means of the analytics cloud, the deployment of the updated model of performed by means of shared object nodes in the *AddressSpace*.

This way, the agents in the field adopt their price during the bidding process according to their recent experiences and with regard to the actual machine failure prediction that they are able to anticipate at any time. This results in an evolving manufacturing system of smart entities that learn on the basis of their internal knowledge, external insights and sensor-based perception.

9 Future Research Topics

9.1 Semantic Scalability for Domain Specific Use-Cases

This work has demonstrated an information modeling approach that is able to describe, integrate and optimize data and information flows within current automation systems of modern factories. As emphasized in the use-case section of this thesis, a decisive factor to strive for benefits of such approach consists in the incorporation of domain-specific knowledge into the development of automated process.

The two application domains that have been examined – namely the *myJoghurt* use-case as well as a manufacturing application – are of completely different nature. However, both processes and their explicit optimization goals and strategies can be pursued by similar approaches. These approaches involve object-oriented modeling techniques, semantic descriptions to provide an explicit representation of domain-specific knowledge as well as a model-based representation of resources. The semantic scalability of these methods allows for an arbitrary recombination and hierarchical structuring of the underlying information models. By making use of this scalability, it becomes possible to manage and to control highly complex processes and to integrate these processes with historical insights and expert knowledge.

To use these scalability concepts in practical applications and real-world systems, the following steps need to be pursued in future factories:

1. Analysis of the current situation in existing production systems.

2. Incorporation of field level devices and field bus systems by common wrapping technologies for enabling OPC UA based communication.

3. Usage of OPC UA companion standards to provide a general framework for a semantic allocation of all field level devices and higher system:

   - Implementation of general companion standards on top of the OPC UA base model such as specifications for *AutomationML* and *ISA 95*.

© Springer Fachmedien Wiesbaden GmbH, part of Springer Nature 2019
M. Hoffmann, *Smart Agents for the Industry 4.0*,
https://doi.org/10.1007/978-3-658-27742-0_9

- Complementation of models by vendor-specific companion specifications, e.g. standards to incorporate IEC 61131-3 capable PLCs.

4. Structuring of all available devices and resources according to these model definitions.

5. Derivation of own companion standards that fit to the manufacturing domain, in which the existing factory is located.

The last step of this process finally enables semantic scalability beyond the basic and standard information model. For the last, which is the most essential one for optimizing an existing manufacturing sites, cooperations with existing OPC UA specification groups or vendors that focus on devices of the same application domain could be very useful. An establishment of overall information models, e.g. for company domain models, can be also useful as a strategic investment in long-term communication solutions for different factories of the same enterprise.

Future research topics will have the main objective to carry out an follow modeling design patterns as indicated above. The *modeling* of a factory in terms of information technological representation has to be equally important as object-oriented software construction for developers. A generic object-oriented approach to these design patterns can be carried out through OPC UA. Together with the according representation of intelligent software agents, this framework can be a powerful toolbox for future factories.

9.2 Generic Extensibility of Communication Concepts Based on Ontologies

Another future-oriented field of research in the target domain directly builds upon the concept of semantic scalability of domain-specific use-cases. The next consequent step in this research direction consist in attempts to automatically generate OPC UA based information models from semantic descriptions such as ontologies.

These ontologies can be available in different type of representations, e.g. by means of OWL ontologies or knowledge graphs. Other ideas in this direction open up opportunities to automatically read and understand product specifications of sensors or field level devices and to transfer these technical specifications into generally understandable and machine-readable OPC UA information models. A recombination of these automatically generated models with existing companion specifications result in a powerful modeling tool to holistically describe production environments without the risk of incalculable efforts.

The achievements that could be obtained in these fields of research can be combined with the results of this work, e.g. to automatically incorporate agent instances with arbitrary shop floor devices and build up dynamic communication networks between these smart entities.

Summary

This thesis introduces a new approach to tackle the challenges of lacking interoperability between tightly-coupled and loosely-coupled systems in modern factories. The OPC Unified Architecture interface and information modeling meta standard is used in multiple ways to solve these problems in a holistic way. Firstly, the interface solutions of OPC UA are used to integrate arbitrary data from low-level systems seamlessly into top-level systems for planning and optimization. Secondly, the semantic information modeling capabilities of OPC UA are utilized to represent cyber-physical systems on the shop floor in the form of a digital twin. As a result, automation systems on the shop floor, smart entities in the field and high-level production planning system interoperate in open and dynamic optimization loops.

A decisive factor for the successful application of the proposed concept is that OPC UA based information models are able to describe all sorts of data and convert them into useful information – no matter where the data is from or in which granularity or time range the data has been collected. By making use of a generic, object-oriented modeling approach, grown manufacturing systems are brought together with modern information management techniques and domain-specific knowledge. This way, legacy systems such as field buses and Industrial Ethernet solutions can be incorporated within SOA approaches realized by OPC UA based architectures.

The incorporation of high-level knowledge into smart agents on the shop floor is just the next logical step in expanding the desired interoperability to the lowest level of the production. Conventional approaches only address one these demands – interoperability between vertically distributed systems or horizontal interoperability and reactivity on the shop floor. The framework that has been presented in this work achieves both goals at the same time. Agents are able to act autonomously and to cooperate with other agents and automation system horizontally on the shop floor. But, at the same time, these agent are also capable of incorporating high-level systems such as shown in terms of the ERP gateway agent that brings the entire knowledge of an enterprise resource planning system down to chip level.

© Springer Fachmedien Wiesbaden GmbH, part of Springer Nature 2019
M. Hoffmann, *Smart Agents for the Industry 4.0*,
https://doi.org/10.1007/978-3-658-27742-0

By making use of the proposed architecture, intelligent agents on the shop floor gain the abilities to incorporate low-level devices in terms of cyber-physical systems and to exchange highly-granular information from the field with top-level planning systems, while at the same time using information from the management level to actively react on changing conditions. Only the combination of the capabilities – tight-coupling in the shop floor and loose access to enterprise services – enables learning that evolves together with the manufacturing system, but without limitations. These learning capabilities, which are enabled by means of the achieved interoperability, are the true key competence to finally reaching reconfigurability, adaptive behavior and real-time factories of the future.

Bibliography

ABB Group (2010). OPC Unified Architecture (UA) Analyzer Device Integration (ADI) and the Potential Application to NeSSI Systems.

Abele, E., Liebeck, T., and Wörn, A. (2006). Measuring Flexibility in Investment Decisions for Manufacturing Systems. *CIRP Annals - Manufacturing Technology*, Volume 55(Issue 1):433–436, doi:10.1016/S0007-8506(07)60452-1.

acatech (2011). *Cyber-Physical Systems. Driving force for innovation in mobility, health, energy and production.* Tech. Report 2011. Acatech, Munich.

Alpaydin, E. (2010). *Introduction to machine learning.* Adaptive computation and machine learning. MIT Press, Cambridge, Mass, 2nd ed. edition.

Andersch, T., Schulz, K., Fritsch, D., Marquardt, K., Jeschke, S., Meisen, T., Tummel, C., Hoffmann, M., and Richert, A. (2015). *Paradigmenwechsel im deutschen Maschinen- und Anlagenbau: Analyse der Herausforderungen und Chancen unter Verwendung eines innovativen, Big-Data-gestützten Ansatzes.* Andersch-Studienreihe: Branchen im strukturellen Wandel. Andersch AG, Frankfurt am Main.

ARC Advisory Group (2014). *Industrial Ethernet-based Devices Worldwide Outlook. FIVE YEAR MARKET ANALYSIS AND TECHNOLOGY FORECAST THROUGH 2015.* ARC Advisory Group. http://www.arcweb.com/market-studies/pages/industrial-ethernet-devices.aspx.

Auberg, H.-W. and Stöger, G. (2016). OPC UA Pub/Sub Model, Real-Time Requirements and IEEE TSN. https://www.tttech.com/news-events/tech-talk/details/interview-softing/.

Auerbach, T., Beckers, M., Buchholz, G., Eppelt, U., Gloy, Y.-S., Fritz, P., Khawli, T., Kratz, S., Lose, J., Molitor, T., Reßmann, A., Thombansen,

U., Veselovac, D., Willms, K., Gries, T., Michaeli, W., Hopmann, C., Reisgen, U., Schmitt, R., and Klocke, F. (2011). Meta-modeling for Manufacturing Processes. In Jeschke, S., Honghai, H., and Schilberg, D., editors, *Intelligent robotics and applications*, volume Part II of *Lecture notes in computer science Lecture notes in artificial intelligence*, pages 199–209. Springer, Berlin.

Azadeh, A., Saberi, M., Kazem, A., Ebrahimipour, V., Nourmohammadza-deh, A., and Saberi, Z. (2013). A flexible algorithm for fault diagnosis in a centrifugal pump with corrupted data and noise based on ANN and support vector machine with hyper-parameters optimization. *Applied Soft Computing*, 13(3):1478–1485, doi:10.1016/j.asoc.2012.06.020.

Baheti, R. and Gill, H. (2011). Cyber-physical systems. *Impact of Control Technology*, 1(1-6).

Baidya, R. and Ghosh, S. K. (2015). Model for a Predictive Maintenance System Effectiveness Using the Analytical Hierarchy Process as Analytical Tool. *IFAC-PapersOnLine*, 48(3):1463–1468, doi:10.1016/j.ifacol.2015.06.293.

Barbosa, J. (2015). *Self-organized and evolvable holonic architecture for manufacturing control*. Dissertation, Université de Valenciennes et du Hainaut Cambrésis. https://tel.archives-ouvertes.fr/tel-01137643v2/document.

Barbosa, J., Leitão, P., Adam, E., and Trentesaux, D. (2015). Dynamic self-organization in holonic multi-agent manufacturing systems: The ADACOR evolution. *Computers in Industry*, 66:99–111, doi:10.1016/j.compind.2014.10.011.

Bauer, A., Bowden, R., Browne, J., Duggan, J., and Lyons, G. (1991). *Shop Floor Control Systems – From Design to Implementation*. Chapman & Hall, USA.

Bauernhansl, T., ten Hompel, M., and Vogel-Heuser, B., editors (2014). *Industrie 4.0 in Produktion, Automatisierung und Logistik: Anwendung · Technologien · Migration*. Springer, Dordrecht.

Bellifemine, F. L., Caire, G., and Greenwood, D. (2007). *Developing multi-agent systems with JADE*. Wiley series in agent technology. John Wiley, Chichester, England and Hoboken, NJ.

Benkedjouh, T., Medjaher, K., Zerhouni, N., and Rechak, S. (2015). Health assessment and life prediction of cutting tools based on support vector regression. *Journal of Intelligent Manufacturing*, 26(2):213–223, doi:10.1007/s10845-013-0774-6.

Billerbeck, J. D. (22.07.2016). Digitalisierung macht vor keiner Branche halt. *VDI Nachrichten*, 2016(29):5. http://www.vdi-nachrichten.com/Technik-Gesellschaft/Digitalisierung-Branche-halt.

Bolton, W. (2009). *Programmable logic controllers*. Newnes, Amsterdam and Boston, fifth edition edition. http://www.sciencedirect.com/science/book/9780128029299.

Bongaerts, L., Valckenaers, P., van Brussel, H., and Wyns, J. (1995). Schedule Execution For A Holonic Shop Floor Control System. *Proc of the Advanced Summer Institute, NOE for ICIMS*, pages 115–124.

Bongaerts, L., Wyns, J., Detand, A., and van Brussel, H. (1996). Identification of manufacturing holons. *Proc. of the Europ. WS for Agent-Oriented Systems in Manufacturing*, TU Berlin, Daimler-Benz:57–73.

Bony, B., Harnischfeger, M., and Jammes, F. (2011). Convergence of OPC UA and DPWS with a cross-domain data model. In *2011 9th IEEE International Conference on Industrial Informatics*, pages 187–192. IEEE, doi:10.1109/INDIN.2011.6034860.

Booth, D., Haas, H., and McCabe, F. (2004). Web Service Architecture: W3C Working Group Note. https://www.w3.org/TR/ws-arch/.

Botti, V. and Giret, A. (2008). *ANEMONA: A Multi-agent Methodology for Holonic Manufacturing Systems*. Springer series in advanced manufacturing. Springer Verlag London Limited, s.l., 1. aufl. edition. http://site.ebrary.com/lib/alltitles/docDetail.action?docID=10239442.

B&R Automation (2015). TSN – A turbo charge for OPC UA? https://www.br-automation.com/smc/bbe7b6f0f6242a5c964fcf843d6c5279cc85f5ab .pdf

Brecher, C. (2011). *Integrative Produktionstechnik für Hochlohnländer*. Springer-Verlag Berlin Heidelberg, Berlin and Heidelberg.

Brennan, R. W. (2007). Toward Real-Time Distributed Intelligent Control: A Survey of Research Themes and Applications. *IEEE Transactions* on *Systems, Man and Cybernetics, Part C (Applications and Reviews)*, 37(5):744–765, doi:10.1109/TSMCC.2007.900670.

Brennan, R. W., Hall, K., Mařik, V., Maturana, F., and Norrie, D. H. (2003). A Real-Time Interface for Holonic Control Devices. In Goos, G., Hartmanis, J., van Leeuwen, J., Mařík, V., McFarlane, D., and Valckenaers, P., editors, *Holonic and Multi-Agent Systems for Manufacturing*, volume 2744 of *Lecture notes in computer science*, pages 25–34. Springer Berlin Heidelberg, Berlin, Heidelberg.

Brennan, R. W. and Norrie, D. H. (2003). From FMS to HMS. In Deen, S. M., editor, *Agent-Based Manufacturing*, pages 31–49. Springer Berlin Heidelberg, Berlin, Heidelberg.

Brown, N. and Kindel, C. (1998). *Distributed Component Object Model Protocol - DCOM/1.0*. Microsoft Corp.

Brückner, S., Wyns, J., Peeters, P., and Kollingbaum, M. (1998). Designing agents for the manufacturing control. *Proceedings on Artificial Intelligence in Manufacturing Workshop*, pages 40–46.

Buffa, E. S. (1983). *Modern production*. Wiley, New York and Chichester, 7th ed. edition.

Busoniu, L., Babuska, R., and de Schutter, B. (2008). A Comprehensive Survey of Multiagent Reinforcement Learning. *IEEE Transactions on Systems, Man, and Cybernetics, Part C (Applications and Reviews)*, 38(2):156–172, doi:10.1109/TSMCC.2007.913919.

Bussmann, S. (1998). An Agent-Oriented Architecture for Holonic Manufacturing Control. *First Open Workshop IMS Europe, Lausanne, Switzerland*, ESPRIT Working Group on IMS.

Bussmann, S., R., N., and Wooldridge, M. (2001). On the Identification of Agents in the Design of Production Control Systems. In Goos, G., Hartmanis, J., van Leeuwen, J., Ciancarini, P., and Wooldridge, M. J., editors, *Agent-Oriented Software Engineering*, volume 1957 of *Lecture notes in computer science*, pages 141–162. Springer Berlin Heidelberg, Berlin, Heidelberg.

Candido, G., Jammes, F., de Oliveira, J. B., and Colombo, A. W. (2010). SOA at device level in the industrial domain: Assessment of OPC UA and DPWS specifications. In *2010 8th IEEE International Conference on Industrial Informatics (INDIN)*, pages 598–603. doi:10.1109/INDIN.2010.5549676.

Çaydaş, U. and Ekici, S. (2012). Support vector machines models for surface roughness prediction in CNC turning of AISI 304 austenitic stainless steel. *Journal of Intelligent Manufacturing*, 23(3):639–650, doi:10.1007/s10845-010-0415-2.

Choi, K., Singh, S., Kodali, A., Pattipati, K. R., Sheppard, J. W., Namburu, S. M., Chigusa, S., Prokhorov, D. V., and Qiao, L. (2009). Novel Classifier Fusion Approaches for Fault Diagnosis in Automotive Systems. *IEEE Transactions on Instrumentation and Measurement*, 58(3):602–611, doi:10.1109/TIM.2008.2004340.

Christensen, J. H. (1994). Holonic manufacturing systems: Initial architecture and standards directions. *First European Conference on Holonic Manufacturing Systems*, Hannover.

Cisek, R., Habicht, C., and Neise, P. (2002). Gestaltung wandlungsfähiger Produktionssysteme. *ZWF - Zeitschrift für wirtschaftlichen Fabrikbetrieb*, 97(9):441–445.

CODESYS (2017). CODESYS - die übergreifende Software-Suite für die Automatisierungstechnik. https://de.codesys.com/das-system.html.

Colombo, A. W. and Karnouskos, S. (2009). Towards the factory of the future: A service-oriented cross-layer infrastructure. In *ICT shaping the world*, ETSI world class standards, pages 65–81. Wiley, Chichester.

Colombo, A. W., Karnouskos, S., Mendes, J. M., and Leitão, P. (2015). Industrial Agents in the Era of Service-Oriented Architectures and Cloud-Based Industrial Infrastructures. In *Industrial Agents*, pages 67–87. Elsevier.

Corsaro, A. (2016). OpenSplice DDS: The Open Source Middleware Accelerating Wall Street. http://www.slideshare.net/Angelo.Corsaro/opensplice-dds-the-open-source-middleware-accelerating-wall-street-1336169.

Cristalli, C., Foehr, M., Jager, T., Leitao, P., Paone, N., Castellini, P., Turrin, C., and Schjolberg, I. (2013). Integration of process and quality control using multi-agent technology. In *2013 IEEE International Symposium on Industrial Electronics*, pages 1–6. IEEE, doi:10.1109/ISIE.2013.6563737.

Crumley, C. L. (1995). Heterarchy and the Analysis of Complex Societies. *Archeological Papers of the American Anthropological Association*, 6(1):1–5, doi:10.1525/ap3a.1995.6.1.1.

Cupek, R., Ziebinski, A., Huczala, L., and Bregulla, M. (2015). Object-Oriented Communication Model for an Agent-Based Inventory Operations Management. *The Fourth International Conference on Intelligent Systems and Applications, The International Symposium on Intelligent Manufacturing Environments (INTELLI 2015)*, pages 80–85.

Cyber Security Dictionary (2012). SCADA. http://www.projectauditors. com/Dictionary2/1.8/index.php/term/,62555c9cae535a6f68555cad5d56. xhtml.

Damba, A. and Watanabe, S. (2007). Hierarchical Control in a Multiagent System. In *Second International Conference on Innovative Computing, Informatio and Control (ICICIC 2007)*, page 111. IEEE, doi:10.1109/ICICIC.2007.334.

Damm, M. (20.05.2015). OPC UA Technical Update.

Damm, M. (2014). OPC UA Discovery.

Danielis, P., Skodzik, J., Altmann, V., Schweissguth, E. B., Golatowski, F., Timmermann, D., and Schacht, J. (2014). Survey on real-time communication via ethernet in industrial automation environments. In *IEEE Emerging Technology and Factory Automation (ETFA 2014)*, pages 1–8. doi:10.1109/ETFA.2014.7005074.

Decotignie, J.-D. (2005). Ethernet-Based Real-Time and Industrial Communications. *Proceedings of the IEEE*, 93(6):1102–1117, doi:10.1109/JPROC.2005.849721.

Detand, A. (1993). *A Computer Aided Process Planning System Generating Non-Linear Process Plans*. PhD thesis, Katholieke Universiteit Leuven, Leuven.

Dictionary.com (2011). *Kanban: Random House Dictionary*. Dictionary.com.

Dietrich, D. and Sauter, T. (2000). Evolution potentials for fieldbus systems. In *2000 IEEE International Workshop on Factory Communication Systems. Proceedings (Cat. No.00TH8531)*, page 343. IEEE, doi:10.1109/WFCS.2000.882567.

Dilts, D. M., Boyd, N. P., and Whorms, H. H. (1991). The evolution of control architectures for automated manufacturing systems. *Journal of Manufacturing Systems*, 10(1):79–93, doi:10.1016/0278-6125(91)90049-8.

DIN 16484-5 (2011). Building automation and control systems (BACS) - Part 5: Data communication protocol.

DIN EN (2010). Functional safety of electrical/electronic/programmable electronic safety-related systems - Part 1: General requirements (IEC 61508-1:2010).

DIN EN 61131-3 (2014). Speicherprogrammierbare Steuerungen.

DIN EN 61158-1 (Feb 2015). Industrielle Kommunikationsnetze – Feldbusse – Teil 1: Überblick und Leitfaden zu den Normen der Reihe IEC 61158 und IEC 61784.

DIN IEC 61987-11 (2010). Industrial-Process Measurement and Control - Data Structures and Elements in Process Equipment Catalogues - Part 11: List of Properties (LOP) of measuring equipment for electronic data exchange - generic structures.

Dong, J., Chen, S., and Jeng, J.-J. (2005). Event-based blackboard architecture for multi-agent systems. In *International Conference on Information Technology: Coding and Computing (ITCC'05) - Volume II*, pages 379–384 Vol. 2. IEEE, doi:10.1109/ITCC.2005.149.

Dong, J., Verhaegen, M., and Gustafsson, F. (2012). Robust Fault Detection With Statistical Uncertainty in Identified Parameters. *IEEE Transactions on Signal Processing*, 60(10):5064–5076, doi:10.1109/TSP.2012.2208638.

ElMaraghy, H., AlGeddawy, T., Azab, A., and ElMaraghy, W. (2012). Change in Manufacturing – Research and Industrial Challenges. In ElMaraghy, H. A., editor, *Enabling Manufacturing Competitiveness and Economic Sustainability*, pages 2–9. Springer Berlin Heidelberg, Berlin, Heidelberg.

ElMaraghy, H. A. (2005). Flexible and reconfigurable manufacturing systems paradigms. *International Journal of Flexible Manufacturing Systems*, 17(4):261–276, doi:10.1007/s10696-006-9028-7.

Epple, U. (2011). Merkmale als Grundlage der Interoperabilität technischer Systeme: Characteristics as Base of System Interoperability. *at - Automatisierungstechnik*, 59(7):440–450.

European Committee for Electrotechnical Standardization (2012). *Smart grids*. CENELEC. https://www.cenelec.eu/aboutcenelec/whatwedo/tech nologysectors/smartgrids.html.

Evans, P. C. and Annunziata, M. (2012). Industrial Internet: Pushing the Boundaries of Minds and Machines.

Fandel, G., Fistek, A., and Stütz, S. (2011). *Produktionsmanagement*. Springer-Lehrbuch. Springer, Berlin and Heidelberg, 2., überarb. und erw. aufl. edition.

Farid, A. M. and Ribeiro, L. (2015). An Axiomatic Design of a Multiagent Reconfigurable Mechatronic System Architecture. *IEEE Transactions on Industrial Informatics*, 11(5):1142–1155, doi:10.1109/TII.2015.2470528.

Fayyad, U. M., editor (1996). *Advances in knowledge discovery and data mining*. AAAI Press, Menlo Park, Calif.

FDI Cooperation (2011). FDI: Field Device Integration Technology.

Feld, J. (2004). PROFINET - scalable factory communication for all applications. In *2004 IEEE International Workshop on Factory Communication Systems*, pages 33–38. doi:10.1109/WFCS.2004.1377673.

Felser, M. (2002). The fieldbus standards: History and structures. *Technology Leadership Day 2002, Organised by MICROSWISS Network*.

Felser, M. (2005). Real-Time Ethernet - Industry Prospective. *Proceedings of the IEEE*, 93(6):1118–1129, doi:10.1109/JPROC.2005.849720.

Fensel, D., Kerrigan, M., and Zaremba, M. (2008). *Implementing Semantic Web services: The SESA framework*. Springer, Berlin.

Fischer, K. (1998). An Agent-Based Approach to Holonic Manufacturing Systems. In Camarinha-Matos, L. M. and Marik, V., editors, *Intelligent Systems for Manufacturing*, pages 3–12. Kluwer Academic Publishers, The Netherlands.

Fletcher, M. and Brennan, R. W. (2001). Designing holonic manufacturing systems using the IEC 61499 (function block) architecture. *IEICE/IEEE Joint Special Issue on Autonomous Decentralized Systems and Systems' Assurance*, E84-D(10):1398–1401.

Fletcher, M., Garcia-Herreros, E., Christensen, J. H., Deen, S. M., and Mittmann, R. (2000). An open architecture for holonic cooperation and autonomy. In *11th International Workshop on Database and Expert Systems Applications*, pages 224–230. doi:10.1109/DEXA.2000.875031.

Foundation for Intelligent Physical Agents (2002a). FIPA ACL Message Structure Specification.

Foundation for Intelligent Physical Agents (2002b). FIPA Agent Management Specification.

Fröhlich, P., Boiger, C., and Kleineberg, O. (2013). Deep Impact: Echtzeit-Ethernet - Neu definiert. http://www.all-electronics.de/deep-impact/.

Gaj, P., Jasperneite, J., and Felser, M. (2013). Computer Communication Within Industrial Distributed Environment—a Survey. *IEEE Transactions on Industrial Informatics*, 9(1):182–189, doi:10.1109/TII.2012.2209668.

Gartner (2015). Gartner's 2015 Hype Cycle for Emerging Technologies Identifies the Computing Innovations That Organizations Should Monitor. http://www.gartner.com/newsroom/id/3114217.

Gausemeier, J., Kahl, S., and Pook, S. (2008a). From mechatronics to self-optimizing systems. In Gausemeier, J., Rammig, F., and Schäfer, W., editors, *Self-optimizing Mechatronic Systems*, pages 3–32. Paderborn.

Gausemeier, J., Rammig, F., and Schäfer, W., editors (2008b). *Self-optimizing Mechatronic Systems: Design The Future*. Paderborn.

Gorjian, N., Ma, L., Mittinty, M., Yarlagadda, P., and Sun, Y. (2010). A review on degradation models in reliability analysis. In Kiritsis, D., Emmanouilidis, C., Koronios, A., and Mathew, J., editors, *Engineering Asset Lifecycle Management*, pages 369–384. Springer London, London.

Grauer, M., Seeger, B., Metz, D., Karadgi, S., and Schneider, M. (2011). About Adopting Event Processing in Manufacturing. In Cezon, M. and Wolfsthal, Y., editors, *Towards a Service-Based Internet. ServiceWave*

2010 Workshops, volume 6569 of *Lecture notes in computer science*, pages 180–187. Springer Berlin Heidelberg, Berlin, Heidelberg.

Groover, M. P. (2013). *Fundamentals of modern manufacturing: Materials, processes, and systems.* John Wiley & Sons, Inc, Hoboken, NJ, 5th ed. edition.

Günther, H.-O. and Tempelmeier, H. (2012). *Produktion und Logistik.* Springer-Lehrbuch. Springer, Berlin and Heidelberg, 9., aktualisierte und erw. aufl. edition.

Günther, J., Pilarski, P. M., Helfrich, G., Shen, H., and Diepold, K. (2014). First Steps Towards an Intelligent Laser Welding Architecture Using Deep Neural Networks and Reinforcement Learning. *Procedia Technology*, 15:474–483, doi:10.1016/j.protcy.2014.09.007.

Guo, X., Sun, L., Li, G., and Wang, S. (2008). A hybrid wavelet analysis and support vector machines in forecasting development of manufacturing. *Expert Systems with Applications*, 35(1-2):415–422, doi:10.1016/j.eswa.2007.07.052.

Hannelius, T., Salmenpera, M., and Kuikka, S. (2008). Roadmap to adopting OPC UA. *6th IEEE International Conference on Industrial Informatics*, pages 756–761, doi:10.1109/INDIN.2008.4618203.

Hannelius, T., Shroff, M., and Tuominen, P. (2009). Embedding OPC Unified Architecture. *Automaatio XVIII Seminari*, Helsinki.

Harding, J. A., Shahbaz, M., Srinivas, and Kusiak, A. (2006). Data Mining in Manufacturing: A Review. *Journal of Manufacturing Science and Engineering*, 128(4):969, doi:10.1115/1.2194554.

Hausmann, S., Gallinat, M., Köster, M., and Heiss, S. (19.01.2015). OPC UA: Ein kritischer Vergleich der IT-Sicherheitsoptionen.

Heimes, F. O. (2008). Recurrent neural networks for remaining useful life estimation. In *2008 International Conference on Prognostics and Health Management*, pages 1–6. IEEE, doi:10.1109/PHM.2008.4711422.

Henßen, R. and Schleipen, M. (2014). Interoperability between OPC UA and AutomationML. *Procedia CIRP*, 25:297–304, doi:10.1016/j.procir.2014.10.042.

Higgins, P., Le Roy, P., and Tierney, L. (1996). *Manufacturing planning and control: Beyond MRP II / Paul Higgins, Patrick Le Roy and Liam Tierney.* Chapman & Hall, London.

Hill, T. (2000). *Manufacturing Strategy: Text and cases.* McGraw Hill, 3rd edition.

Hochreiter, S. and Schmidhuber, J. (1997). Long Short-Term Memory. *Neural Computation*, 9(8):1735–1780, doi:10.1162/neco.1997.9.8.1735.

Hoffmann, M., Büscher, C., Meisen, T., and Jeschke, S. (2016a). Continuous Integration of Field Level Production Data into Top-level Information Systems Using the OPC Interface Standard. *Procedia CIRP*, 41:496–501, doi:10.1016/j.procir.2015.12.059.

Hoffmann, M., Büscher, C., Meisen, T., and Jeschke, S. (2016b). OPC UA basierte Multi-Agenten-Systeme: Intelligente Automatisierung gewachsener Produktionsanlagen. *atp - Automatisierungstechnische Praxis*, (Hauptbeitrag)(7-8):54–65.

Hoffmann, M., Büscher, C., Meisen, T., and Jeschke, S. (2016c). Sichere und zuverlässige Integration von Multi-Agenten-Systemen und Cyber-Physischen Systemen für eine intelligente, dynamische Produktionssteuerung auf Basis von OPC UA. *Automation 2016: secure & reliable in the digital world: 17. Branchentreff der Mess- und Automatisierungstechnik,* Baden-Baden, 07. und 08. Juni 2016 / VDI-Berichte 2284, Düsseldorf: VDI Verlag GmbH:119–120.

Hoffmann, M., Kreisköther, K., Büscher, C., Meisen, T., Kampker, A., Schilberg, D., and Jeschke, S. (2014). Optimized Factory Planning and Process Chain Formation using Virtual Production Intelligence. *Enabling manufacturing competitiveness and economic sustainability - Proceedings of the 5th International Conference on Changeable, Agile, Reconfigurable and Virtual Production (CARV 2013).*, Munich, Germany, October 6th-9th:153–158, doi:10.1007/978-3-319-02054-9_26.

Hoffmann, M., Rix, M., Büscher, C., Sauber, K., and Meisen, T. (2016d). Integration von Industrie 4.0-Lösungen in bestehende Produktionsanlagen: Erfolgreiche Umsetzung durch ein Open-Source-Framework. *ProductivITy*, 12(2):16–18.

Hoffmann, M., Thomas, P., Schutz, D., Vogel-Heuser, B., Meisen, T., and Jeschke, S. (2016e). Semantic integration of multi-agent systems using an OPC UA information modeling approach. In *2016 IEEE 14th International Conference on Industrial Informatics (INDIN)*, pages 744–747. IEEE, doi:10.1109/INDIN.2016.7819258.

Hoppe, S. (2014). Standardisierte horizontale und vertikale Kommunikation: Status und Ausblick. In Bauernhansl, T., ten Hompel, M., and Vogel-Heuser, B., editors, *Industrie 4.0 in Produktion, Automatisierung und Logistik*, pages 325–341. Springer, Dordrecht.

Huhns, M. N., editor (2000). *Readings in agents*. Kaufmann, San Francisco, Calif., [3. print.] edition.

Huhns, M. N. and Singh, M. P. (2005). Service-oriented computing: Key concepts and principles. *IEEE Internet Computing*, 9(1):75–81, doi:10.1109/MIC.2005.21.

IEC 61499 (2000). IEC 61499. Function Block: Part 1 - Architecture. PAS 61499-1.

IEC 61499 (2001). IEC 61499. Function Block: Part 2 - Software Tool Requirements. PAS 61499-2.

IEC 61804 (2015). Function blocks (FB) for process control and electronic device description language (EDDL).

IEC 61850 (2013). Power Utility Automation.

IEC 62541 (2015). OPC Unified Architecture. https://opcfoundation.org/ developer-tools/ specifications-unified-architecture.

IEC 62541-1 (2015). OPC Unified Architecture - Part 1 - Overview and Concepts. https://opcfoundation.org/developer-tools/specifications-unified-architecture.

IEC 62541-10 (2015). OPC Unified Architecture - Part 10 - Programs. https://opcfoundation.org/developer-tools/specifications-unified-architecture.

IEC 62541-11 (2015). OPC Unified Architecture - Part 11 - Historical Access. https://opcfoundation.org/developer-tools/specifications-unified-architecture.

IEC 62541-12 (2015). OPC Unified Architecture - Part 12 - Discovery. https://opcfoundation.org/developer-tools/specifications-unified-architecture.

IEC 62541-13 (2015). OPC Unified Architecture - Part 13 - Aggregates. https://opcfoundation.org/developer-tools/specifications-unified-architecture.

IEC 62541-2 (2015). OPC Unified Architecture - Part 2 - Security Model. https://opcfoundation.org/developer-tools/specifications-unified-architecture.

IEC 62541-3 (2015). OPC Unified Architecture - Part 3 - Address Space Model. https://opcfoundation.org/developer-tools/specifications-unified-architecture.

IEC 62541-4 (2015). OPC Unified Architecture - Part 4 - Services. https://opcfoundation.org/developer-tools/specifications-unified-architecture.

IEC 62541-5 (2015). OPC Unified Architecture - Part 5 - Information Model. https://opcfoundation.org/developer-tools/specifications-unified-architecture.

IEC 62541-6 (2015). OPC Unified Architecture - Part 6 - Mappings. https://opcfoundation.org/developer-tools/specifications-unified-architecture.

IEC 62541-7 (2015). OPC Unified Architecture - Part 7 - Profiles. https://opcfoundation.org/developer-tools/specifications-unified-architecture.

IEC 62541-8 (2015). OPC Unified Architecture - Part 8 - Data Access. https://opcfoundation.org/developer-tools/specifications-unified-architecture.

IEC 62541-9 (2015). OPC Unified Architecture - Part 9 - Alarms and Conditions. https://opcfoundation.org/developer-tools/specifications-unified-architecture.

IEEE (1980). IEEE 802.3 - ETHERNET - Institute of Electrical and Electronics Engineers. http://www.ieee802.org/3/.

IEEE (1998). Carrier sense multiple access with collision detection (CSMA/CD) access method and physical layer specifications, Frame Extension for Virtual Bridged Local Area Networks (VLAN) Tagging on 802.3 Networks.

IEEE (2016). IEEE Glossary: Interoperability. http://www.ieee.org/edu cation_careers/education/standards/standards_glossary.html.

IEEE 1722-2011 (2011). 1722-2011 - IEEE Standard for Layer 2 Transport Protocol for Time Sensitive Applications in a Bridged Local Area Network. http://standards.ieee.org/findstds/standard/1722-2011.html.

IEEE 802.1BA (19.07.2011). 802.1BA - Audio Video Bridging (AVB) Systems. http://www.ieee802.org/1/pages/802.1ba.html.

IEEE 802.1Qat (26.06.2010). 802.1Qat - Stream Reservation Protocol. http://www.ieee802.org/1/pages/802.1at.html.

IEEE 802.1Qav (04.11.2009). 802.1Qav - Forwarding and Queuing Enhancements for Time-Sensitive Streams. http://www.ieee802.org/1/pages/802.1av.html.

IEEE 802.1Qbv (07.10.2015). 802.1Qbv - Enhancements for Scheduled Traffic. http://www.ieee802.org/1/pages/802.1bv.html.

International Electrotechnical Commission (2013). IEC 62264 - Enterprise-control system integration.

ISA (2000). ISA-95.

Iwata, K., Onosato, M., and Koike, M. (1994). Random Manufacturing System: A New Concept of Manufacturing Systems for Production to Order. *CIRP Annals - Manufacturing Technology*, 43(1):379–383, doi:10.1016/S0007-8506(07)62235-5.

Iyer, N., Goebel, K., and Bonissone, P. (2006). Framework for Post-Prognostic Decision Support. In *2006 IEEE Aerospace Conference*, pages 1–10. IEEE, doi:10.1109/AERO.2006.1656108.

Izaguirre, M. J. A. G., Lobov, A., and Lastra, J. L. M. (2011). OPC-UA and DPWS interoperability for factory floor monitoring using complex event processing. In *2011 9th IEEE International Conference on Industrial Informatics (INDIN)*, pages 205–211. doi:10.1109/INDIN.2011.6034874.

Jacobs, F. R. and Bragg, D. J. (1988). REPETITIVE LOTS: FLOW-TIME REDUCTIONS THROUGH SEQUENCING AND DYNAMIC BATCH SIZING. *Decision Sciences*, 19(2):281–294, doi:10.1111/j.1540-5915.1988.tb00267.x.

Jammes, F. and Smit, H. (2005). Service-Oriented Paradigms in Industrial Automation. *IEEE Trans. on Industrial Informatics*, 1(1):62–70, doi:10.1109/TII.2005.844419.

Jammes, F., Smit, H., Martinez Lastra, J. L., and Delamer, I. M. (2005). Orchestration of Service-Oriented Manufacturing Processes. In *2005 IEEE Conference on Emerging Technologies and Factory Automation*, pages 617–624. IEEE, doi:10.1109/ETFA.2005.1612580.

Jasperneite, J. and Feld, J. (2005). PROFINET: An Integration Platform for heterogeneous Industrial Communication Systems. In *2005 IEEE Conference on Emerging Technologies and Factory Automation (ETFA 2005)*, pages 815–822. doi:10.1109/ETFA.2005.1612610.

Jasperneite, J. and Neumann, P. (2004). How to guarantee realtime behavior using Ehernet. *11th IFAC Symposium on Information Control Problems in Manufacturing (INCOM'2004)*, Salvador-Bahia, Brazil.

Jazdi, N. (2014). Cyber physical systems in the context of Industry 4.0. In *2014 IEEE International Conference on Automation, Quality and Testing, Robotics*, pages 1–4. IEEE, doi:10.1109/AQTR.2014.6857843.

Jean Vieille (2010). An ISA-95 companion standard for OPC-UA.

Jedrzejowicz, P. (2011). Machine Learning and Agents. In O'Shea, J., editor, *Agent and multi-agent systems technologies and applications*, LNCS sublibrary. SL 7, Artificial intelligence. Springer, Heidelberg.

Jennings, N. R. (2000). On agent-based software engineering. *Artificial Intelligence*, 117(2):277–296, doi:10.1016/S0004-3702(99)00107-1.

Jennings, N. R. and Wooldridge, M. (1998). Applications of Intelligent Agents. In Jennings, N. R. and Wooldridge, M. J., editors, *Agent Technology*, pages 3–28. Springer Berlin Heidelberg, Berlin, Heidelberg.

Jin-Hai, L., Anderson, A. R., and Harrison, R. T. (2003). The evolution of agile manufacturing. *Business Process Management Journal*, 9(2):170–189, doi:10.1108/14637150310468380.

John, D. and Jasperneite, J. (2011). Interoperabilität auf Feldebene. *at - Automatisierungstechnik*, 59(7):406–412.

Kagermann, H., Wahlster, W., and Helbig, J. (2013). Securing the future of German manufacturing industry: Recommendations for implementing the strategic initiative INDUSTRIE 4.0.

Karnouskos, S., Bangemann, T., and Diedrich, C. (2009). Integration of Legacy Devices in the Future SOA-based Factory. *IFAC Proceedings Volumes*, 42(4):2113–2118, doi:10.3182/20090603-3-RU-2001.0487.

Karnouskos, S., Havlena, V., Jerhotova, E., Kodet, P., Sikora, M., Stluka, P., Trnka, P., and Tilly, M. (2014). Plant Energy Management. In Colombo, A. W., Bangemann, T., Karnouskos, S., Delsing, J., Stluka, P., Harrison, R., Jammes, F., and Lastra, J. L., editors, *Industrial Cloud-Based Cyber-Physical Systems*, pages 203–218. Springer International Publishing, Cham.

Karnouskos, S. and Tariq, M. M. J. (2008). An Agent-Based Simulation of SOA-Ready Devices. In *Tenth International Conference on Computer Modeling and Simulation (uksim 2008)*, pages 330–335. IEEE, doi:10.1109/UKSIM.2008.81.

Karnouskos, S. and Tariq, M. M. J. (2009). Using multi-agent systems to simulate dynamic infrastructures populated with large numbers of web service enabled devices. In *2009 International Symposium on Autonomous Decentralized Systems*, pages 1–7. IEEE, doi:10.1109/ISADS.2009.5207354.

Kendall, E. A. and Malkoun, M. T. (1997). Design patterns for the development of multiagent systems. In Carbonell, J. G., Siekmann, J., Goos, G., Hartmanis, J., van Leeuwen, J., Zhang, C., and Lukose, D., editors, *Multi-Agent Systems Methodologies and Applications*, volume 1286 of *Lecture notes in computer science*, pages 17–31. Springer Berlin Heidelberg, Berlin, Heidelberg.

Kimura, F. (1993). Product and Process Modelling as a Kernel for Virtual Manufacturing Environment. *CIRP Annals - Manufacturing Technology*, 42(1):147–150, doi:10.1016/S0007-8506(07)62413-5.

Kletti, J., editor (2007). *Manufacturing Execution Systems - MES*. Springer, Berlin u.a.

Knowles, J. and Nakayama, H. (2008). Meta-Modeling in Multiobjective Optimization. In Branke, J., editor, *Multiobjective Optimization*, volume 5252 of *LNCS sublibrary. Theoretical computer science and general issues*, pages 245–284. Springer-Verlag, Berlin.

Koestler, A. (1989). *The ghost in the machine.* Arkana, London.

Koren, Y., Heisel, U., Jovane, F., Moriwaki, T., Pritschow, G., Ulsoy, G., and van Brussel, H. (1999). Reconfigurable Manufacturing Systems. *CIRP Annals - Manufacturing Technology*, 48(2):527–540, doi:10.1016/S0007-8506(07)63232-6.

Koschnick, G. (2015). Das Referenzarchitekturmodell Industrie 4.0 (RAMI 4.0).

Kumar, P. and Singhal, N. (2014). Web Crawling Using Dynamic IP Address Using Single Server. *International Journal of Advanced Research in Computer and Communication Engineering (IJARCCE)*, Vol. 3(Issue 1).

Kurbel, K. (2016). *De Gruyter Studium: Enterprise Resource Planning und Supply Chain Management in der Industrie : Von MRP bis Industrie 4.0 (8).* De Gruyter Studium. De Gruyter, 0008 fully revised a edition.

Lange, J., Burke, T. J., and Iwanitz, F. (2014). *OPC: Von Data Access bis Unified Architecture.* VDE Verl., Berlin, 5., durchges. aufl edition.

Lange, J., Iwanitz, F., and Burke, T. J. (2010). *OPC: From Data Access to Unified Architecture.* VDE-Verl., Berlin and Offenbach, 4., rev. ed edition.

Lastra, J. (2006). Semantic Web Services in Factory Automation: Fundamental Insights and Research Roadmap. *IEEE Transactions on Industrial Informatics*, 2(1):1–11, doi:10.1109/TII.2005.862144.

Lastra, J. L. M., Torres, E. L., and Colombo, A. W. (2005). A 3D Visualization and Simulation Framework for Intelligent Physical Agents. In Hutchison, D., Kanade, T., Kittler, J., Kleinberg, J. M., Mattern, F., Mitchell, J. C., Naor, M., Nierstrasz, O., Pandu Rangan, C., Steffen, B., Sudan, M., Terzopoulos, D., Tygar, D., Vardi, M. Y., Weikum, G., Mařík, V., William Brennan, R., and Pěchouček, M., editors, *Holonic and Multi-Agent Systems for Manufacturing*, volume 3593 of *Lecture notes in computer science*, pages 23–38. Springer Berlin Heidelberg, Berlin, Heidelberg.

Lee, E. A. (2008). Cyber Physical Systems: Design - Challenges. *Electrical Engineering and Computer Science*, University of Berkeley. Technical Report No. UCB/EECS-2008-8. http://www.eecs.berkeley.edu/Pubs/TechRpts/2008/EECS-2008-8.html.

Lee, J., Bagheri, B., and Kao, H.-A. (2015). A Cyber-Physical Systems architecture for Industry 4.0-based manufacturing systems. *Manufacturing Letters*, 3:18–23, doi:10.1016/j.mfglet.2014.12.001.

Lee, J., Lapira, E., Bagheri, B., and Kao, H.-A. (2013). Recent advances and trends in predictive manufacturing systems in big data environment. *Manufacturing Letters*, 1(1):38–41, doi:10.1016/j.mfglet.2013.09.005.

Legat, C., Schütz, D., and Vogel-Heuser, B. (2014). Automatic generation of field control strategies for supporting (re-)engineering of manufacturing systems. *Journal of Intelligent Manufacturing*, 25(5):1101–1111, doi:10.1007/s10845-013-0744-z.

Leinweber, D. J. (2007). Stupid Data Miner Tricks. *The Journal of Investing*, 16(1):15–22, doi:10.3905/joi.2007.681820.

Leitão, P., Barbosa, J., and Trentesaux, D. (2012). Bio-inspired multi-agent systems for reconfigurable manufacturing systems. *Engineering Applications of Artificial Intelligence*, 25(5):934–944, doi:10.1016/j.engappai.2011.09.025.

Leitão, P. and Karnouskos, S. (2015). *Industrial agents: Emerging applications of software agents in industry*.

Leitão, P., Karnouskos, S., Ribeiro, L., Lee, J., Strasser, T., and Colombo, A. W. (2016). Smart Agents in Industrial Cyber–Physical Systems. *Proceedings of the IEEE*, 104(5):1086–1101, doi:10.1109/JPROC.2016.2521931.

Leitão, P. and Restivo, F. J. (2008). Implementation of a Holonic Control System in a Flexible Manufacturing System. *IEEE Transactions on Systems, Man and Cybernetics, Part C (Applications and Reviews)*, 38(5):699–709, doi:10.1109/TSMCC.2008.923881.

Leitner, S.-H. and Mahnke, W. (2006). OPC UA - Service-oriented Architecture for Industrial Applications. *ABB Corporate Research Center*.

Leone, G. and Rahn, R. D. (2002). *Fundamentals of flow manufacturing.* Flow Pub, Boulder, Colo.

Liker, J. K. (2004). *The Toyota way: 14 management principles from the world's greatest manufacturer.* McGraw-Hill, New York, NY. http://www.loc.gov/catdir/bios/mh041/2004300007.html.

Lin, S.-W. and Miller, B. e. a. (2015). *Industrial Internet Reference Architecture.* Tech. Report 2015. http://www.iiconsortium.org/IIRA-1-7-ajs.pdf.

LoBello, L., Kaczynski, G. A., and Mirabella, O. (2005). Improving the Real-Time Behavior of Ethernet Networks Using Traffic Smoothing. *IEEE Transactions on Industrial Informatics,* 1(3):151–161, doi:10.1109/TII.2005.852071.

Lu, S. C.-Y. (1990). Machine learning approaches to knowledge synthesis and integration tasks for advanced engineering automation. *Computers in Industry,* 15(1-2):105–120, doi:10.1016/0166-3615(90)90088-7.

Luckham, D. C. (2006). What's the Difference Between ESP and CEP? *www.complexevents.com.* http://www.complexevents.com/2006/08/01/what's-the-difference-between-esp-and-cep/.

Lüder, A. (2014). Integration des Menschen in Szenarien der Industrie 4.0. In Bauernhansl, T., ten Hompel, M., and Vogel-Heuser, B., editors, *Industrie 4.0 in Produktion, Automatisierung und Logistik,* pages 493–507. Springer, Dordrecht.

Lüder, A., Peschke, J., Sauter, T., Deter, S., and Diep, D. (2004). Distributed intelligence for plant automation based on multi-agent systems: The PABADIS approach. *Production Planning & Control,* 15(2):201–212, doi:10.1080/09537280410001667484.

Mahnke, W., Damm, M., and Leitner, S.-H. (2009). *OPC unified architecture.* Springer, Berlin [u.a.].

Mahnke, W., Gossling, A., Graube, M., and Urbas, L. (2011). Information modeling for middleware in automation. *16th Conference on Emerging Technologies & Factory Automation (ETFA),* IEEE:1–7.

Maka, A., Cupek, R., and Rosner, J. (2011a). OPC UA Object Oriented Model for Public Transportation System. In *2011 European Modelling Symposium (EMS),* pages 311–316. doi:10.1109/EMS.2011.84.

Maka, A., Cupek, R., and Wierzchanowski, M. (2011b). Agent-based Modeling for Warehouse Logistics Systems. In *2011 UkSim 13th International Conference on Computer Modelling and Simulation*, pages 151–155. IEEE, doi:10.1109/UKSIM.2011.37.

Malakooti, B. (2014). *Production and operation systems with multiple objectives*. Wiley series in systems engineering and management. Wiley, Hoboken, New Jersey. http://site.ebrary.com/lib/alltitles/docDetail.action?docID=10814677.

Mařík, V., Vrba, P., Hall, K. H., and Maturana, F. P. (2005). Rockwell automation agents for manufacturing. In Pechoucek, M., Steiner, D., and Thompson, S., editors, *the fourth international joint conference*, page 107. doi:10.1145/1082473.1082812.

Marin, C. A., Monch, L., Leitao, P., Vrba, P., Kazanskaia, D., Chepegin, V., Liu, L., and Mehandjiev, N. (2013). A Conceptual Architecture Based on Intelligent Services for Manufacturing Support Systems. In *2013 IEEE International Conference on Systems, Man, and Cybernetics*, pages 4749–4754. IEEE, doi:10.1109/SMC.2013.808.

Mayer, F. (2014). Austauschspezifikation v2: JPP.

Mayer, F., Pantförder, D., Diedrich, C., and Vogel-Heuser, B. (2013). Deutschlandweiter I4.0-Demonstrator: Technisches Konzept und Implementierung. http://nbn-resolving.de/urn/resolver.pl?urn:nbn:de:bvb:91-epub-20131112-1178726-0-0.

McClellan, M. (1997). *Applying manufacturing execution systems*. The St. Lucie Press/APICS series on resource management. St. Lucie Press and APICS, Boca Raton, Fla. and Falls Church, Va. http://www.loc.gov/catdir/enhancements/fy0646/97204570-d.html.

McClellan, M. (2001). Introduction to Manufacturing Execution Systems. *MES Conference & Exposition 2001*, Baltimore, Maryland.

McFarlane, D. (1995). Holonic manufacturing systems in continuous processing: concepts and control requirements. *Adv. Summer Inst. (ASI'95) on Life Cycle Approaches to Production Systems*, Lisbon, Portugal.

McFarlane, D. C. and Bussmann, S. (2003). Holonic Manufacturing Control: Rationales, Developments and Open Issues. In Deen, S. M., editor, *Agent-*

Based Manufacturing, pages 303–326. Springer Berlin Heidelberg, Berlin, Heidelberg.

Medjaher, K., Tobon-Mejia, D. A., and Zerhouni, N. (2012). Remaining Useful Life Estimation of Critical Components With Application to Bearings. *IEEE Transactions on Reliability*, 61(2):292–302, doi:10.1109/TR.2012.2194175.

Meisen, T., Rix, M., Hoffmann, M., Schilberg, D., and Jeschke, S. (2013). A Framework for Semantic Integration and Analysis of Measurement Data in Modern Industrial Machinery. *11th International Symposium of Measurement Technology and Intelligent Instruments*.

Mendes, J. M., Leitao, P., Restivo, F., and Colombo, A. W. (2009). Service-Oriented Agents for Collaborative Industrial Automation and Production Systems. In Mařík, V., Strasser, T., and Zoitl, A., editors, *Holonic and multi-agent systems for manufacturing*, Lecture notes in computer science Lecture notes in artificial intelligence. Springer, Berlin.

Metz, D., editor (2014). *The Concept of a Real-Time Enterprise in Manufacturing*. Springer Fachmedien Wiesbaden, Wiesbaden. doi:10.1007/978-3-658-03750-5.

Microsoft (12.01.2017). Unterstützung für AMQP 1.0 in Service Bus. https://docs.microsoft.com/de-de/azure/service-bus-messaging/service-bus-amqp-overview.

Minor, J. (2011). *Bridging OPC UA and DPWS for industrial SOA*. Dissertation, Tampere University of Technology. http://dspace.cc.tut.fi/dpub/bitstream/handle/123456789/20954/minor.pdf.

Mirshekarian, S. and Šormaz, D. N. (2016). Correlation of job-shop scheduling problem features with scheduling efficiency. *Expert Systems with Applications*, 62:131–147, doi:10.1016/j.eswa.2016.06.014.

Mönch, L., Stehli, M., and Zimmermann, J. (2003). FABMAS: An Agent-Based System for Production Control of Semiconductor Manufacturing Processes. In Goos, G., Hartmanis, J., van Leeuwen, J., Mařík, V., McFarlane, D., and Valckenaers, P., editors, *Holonic and Multi-Agent Systems for Manufacturing*, volume 2744 of *Lecture notes in computer science*, pages 258–267. Springer Berlin Heidelberg, Berlin, Heidelberg.

Monostori, L. (2003). AI and machine learning techniques for managing complexity, changes and uncertainties in manufacturing. *Engineering Applications of Artificial Intelligence*, 16(4):277–291, doi:10.1016/S0952-1976(03)00078-2.

Mosallam, A., Medjaher, K., and Zerhouni, N. (2016). Data-driven prognostic method based on Bayesian approaches for direct remaining useful life prediction. *Journal of Intelligent Manufacturing*, 27(5):1037–1048, doi:10.1007/s10845-014-0933-4.

MTConnect Institute (November 20th, 2013). MTConnect OPC UA Companion Specification.

Müller, J. P. (1996). The design of intelligent agents: a layered approach. *LNAI 1177*, Springer-Verlag.

Müller, J. P. and Pischel, M. (1993). The agent architecture InteRRaP : concept and application. (93-26).

Najid, N. M., Kouiss, K., and Derriche, O. (2002). Agent based approach for a real-time shop floor control. In *IEEE International Conference on Systems, Man and Cybernetics*, page 6. IEEE, doi:10.1109/ICSMC.2002.1173292.

NIST (2010). *NIST Framework and Roadmap for Smart Grid Interoperability Standards*. NIST.

OASIS (29 October 2014). MQTT. http://docs.oasis-open.org/mqtt/mqtt/v3.1.1/os/mqtt-v3.1.1-os.pdf.

Obermeier, D. (2015). MQTT: Schnelleinstieg in das schlanke IoT-Protokoll mit Java: IoT-Allrounder. https://jaxenter.de/iot-allrounder-27208.

O'Hare, G. M. P. and Jennings, N. R., editors (1996). *Foundations of distributed artificial intelligence*. 6th-generation comp. tech. ser.. Wiley, New York. http://www.loc.gov/catdir/bios/wiley041/95000238.html.

Okino, N. (1993). Bionic Manufacturing Systems. In Peklenik, J., editor, *CIRP Flexible Manufacturing Systems Past-Present.Future*, pages 73–95.

Öno, T. and Bodek, N. (2008). *Toyota production system: Beyond large-scale production*. Productivity Press, New York, NY, [reprinted] edition.

Osterhage, W. W. (2014). *ERP-Kompendium: Eine Evaluierung von Enterprise Resource Planning Systemen.* Xpert.press. Springer Berlin Heidelberg, Berlin, Heidelberg, aufl. 2014 edition.

Otto, A. and Hellmann, K. (2009). IEC 61131: A general overview and emerging trends. *IEEE Industrial Electronics Magazine*, 3(4):27–31, doi:10.1109/MIE.2009.934793.

Pantförder, D., Mayer, F., Diedrich, C., Göhner, P., Weyrich, M., and Vogel-Heuser, B. (2014). Agentenbasierte dynamische Rekonfiguration von vernetzten intelligenten Produktionsanlagen – Evolution statt Revolution. In Bauernhansl, T., ten Hompel, M., and Vogel-Heuser, B., editors, *Industrie 4.0 in Produktion, Automatisierung und Logistik*, pages 145–158. Springer, Dordrecht.

Pardo-Castellote, G., Farabaugh, B., and Warren, R. (2005). An Introduction to DDS and Data-Centric Communications.

Peltz, C. (2003). Web services orchestration and choreography. *Computer*, 36(10):46–52, doi:10.1109/MC.2003.1236471.

Permin, E., Hoffmann, M., Bertelsmeier, F., Haag, S., Detert, T., and Schmitt, R. (2015). Cognitive Self-Optimization in Industrial Assembly. In Wulfsberg, J. P., Röhlig, B., and Montag, T., editors, *Progress in Production Engineering*, Applied Mechanics and Materials. Trans Tech Publications Inc., Pfaffikon.

Pessemier, W., Deconinck, G., Raskin, G., Saey, P., and van Winckel, H. (2011). Suitability assessment of OPC UA as the backbone of ground-based observatory control systems. *13th International Conference on Accelerator and Large Experimental Physics Control Systems (ICALEPCS)*, pages 1174–1177.

Phoenix Contact (2005). Industrial Ethernet Products. ethernetrail.com.

Pinedo, M. (2014). *Scheduling: Theory, algorithms, and systems.* Springer, New York, 4th edition edition.

Piontek, J. (2005). *Controlling.* Managementwissen für Studium und Praxis. Oldenbourg, München, 3., erw. aufl. edition. http://www.oldenbourg-link.com/doi/book/10.1524/9783486700350.

Pöschmann, A. (2001). Chancen und Nutzen von Ethernet TCP/IP in der Automatisierung. *GMA-Kongress 2001: Automatisierungstechnik im Spannungsfeld neuer Technologien*, VDI-Berichte Nr. 1608:617–626.

Prytz, G. (2008). A performance analysis of EtherCAT and PROFINET IRT. In *2008 IEEE International Conference on Emerging Technologies and Factory Automation (ETFA 2008)*, pages 408–415. doi:10.1109/ETFA.2008.4638425.

Real-Time Innovations (13.04.2016). Object Management Group and OPC Foundation Announce Collaborative Strategy for the DDS and OPC UA Connectivity Standards: The Two Standards Organizations, along with the Industrial Internet Consortium (IIC), Industrie 4.0, and leading DDS and OPC UA Vendors Enable Immediate Industrial IoT Market Adoption. http://news.rti.com/pr/omg-opcf-collaborative-strategy.

Resnick, C. (2016). MQTT is Gaining Ground as a Messaging Protocol for IIoT. https://industrial-iot.com/2016/10/mqtt-protocol-for-iiot/.

Ribeiro, B. (2005). Support Vector Machines for Quality Monitoring in a Plastic Injection Molding Process. *IEEE Transactions on Systems, Man and Cybernetics, Part C (Applications and Reviews)*, 35(3):401–410, doi:10.1109/TSMCC.2004.843228.

Ribeiro, L., Barata, J., and Mendes, P. (2008). MAS and SOA: Complementary Automation Paradigms. In Azevedo, A., editor, *Innovation in Manufacturing Networks*, volume 266 of *IFIP – The International Federation for Information Processing*, pages 259–268. Springer US, Boston, MA.

Ribeiro, L., Barata, J., Onori, M., and Hoos, J. (2015). Industrial Agents for the Fast Deployment of Evolvable Assembly Systems. In *Industrial Agents*, pages 301–322. Elsevier.

Ricot, C. (2001). Challenges and opportunities in industrial automation. *8th IEEE International Conference on Emerging Technologies and Factory Automation*.

Rinaldi, J. (2013). *OPC UA - the basics: An OPC UA overview for those who are not networking gurus*. Amazon, Great Britain.

Rocha, A., Di Orio, G., Barata, J., Antzoulatos, N., Castro, E., Scrimieri, D., Ratchev, S., and Ribeiro, L. (2014). An agent based framework to support plug and produce. In *2014 12th IEEE International Conference on Industrial Informatics (INDIN)*, pages 504–510. IEEE, doi:10.1109/INDIN.2014.6945565.

Rohjans, S., Fensel, D., and Fensel, A. (2011). OPC UA goes semantics: Integrated communications in smart grids. In *Factory Automation (ETFA 2011)*, pages 1–4. doi:10.1109/ETFA.2011.6059133.

Rohjans, S., Uslar, M., and Appelrath, H. J. (2010). OPC UA and CIM: Semantics for the smart grid. In *2010 IEEE PES Transmission & Distribution Conference & Exposition*, pages 1–8. IEEE, [Piscataway, N.J.].

Russell, S., Norvig, P., and Davis, E. (2010). *Artificial intelligence: A modern approach*. Pearson, Boston, 3rd ed. edition.

Russell, S. J. (2016). *Artificial intelligence: A modern approach*. Prentice Hall series in artificial intelligence. Pearson Education, Boston and Columbus and Indianapolis and New York and San Francisco, global ed. of 3rd revised ed. edition.

Salahshoor, K., Kordestani, M., and Khoshro, M. S. (2010). Fault detection and diagnosis of an industrial steam turbine using fusion of SVM (support vector machine) and ANFIS (adaptive neuro-fuzzy inference system) classifiers. *Energy*, 35(12):5472–5482, doi:10.1016/j.energy.2010.06.001.

Sardinha, J. A. R. P., Garcia, A., Lucena, C. J. P., and Milidiú, R. L. (2005). A Systematic Approach for Including Machine Learning in Multi-agent Systems. In Hutchison, D., Kanade, T., Kittler, J., Kleinberg, J. M., Mattern, F., Mitchell, J. C., Naor, M., Nierstrasz, O., Pandu Rangan, C., Steffen, B., Sudan, M., Terzopoulos, D., Tygar, D., Vardi, M. Y., Weikum, G., Bresciani, P., Giorgini, P., Henderson-Sellers, B., Low, G., and Winikoff, M., editors, *Agent-Oriented Information Systems II*, volume 3508 of *Lecture notes in computer science*, pages 198–211. Springer Berlin Heidelberg, Berlin, Heidelberg.

Sauter, T. (2005). Fieldbus systems – History and Evolution. In Zurawski, R., editor, *The industrial communication technology handbook*, volume Chapter 7 of *Industrial information technology series*. CRC Press.

Sauter, T. (2007). The continuing evolution of integration in manufacturing automation. *IEEE Industrial Electronics Magazine*, 1(1):10–19, doi:10.1109/MIE.2007.357183.

Sauter, T. (2010). The Three Generations of Field-Level Networks—Evolution and Compatibility Issues. *IEEE Transactions on Industrial Electronics*, 57(11):3585–3595, doi:10.1109/TIE.2010.2062473.

Saxena, A., Goebel, K., Simon, D., and Eklund, N. (2008). Damage propagation modeling for aircraft engine run-to-failure simulation. In *2008 International Conference on Prognostics and Health Management*, pages 1–9. IEEE, doi:10.1109/PHM.2008.4711414.

Schumacher, J. (2015). Industry 4.0 – specifically.

Schumacher, M., Jasperneite, J., and Weber, K. (2008). A new approach for increasing the performance of the industrial Ethernet system PROFINET. In *2008 IEEE International Workshop on Factory Communication Systems - (WFCS 2008)*, pages 159–167. doi:10.1109/WFCS.2008.4638725.

Schütte, S., Nieße, A., Rohjans, S., and Rohlfs, H. (2013). OPC UA Compliant Coupling of Multi-Agent Systems and Smart Grid Simulations. In *IECON 2013-39th Annual Conference of the IEEE Industrial Electronics Society*, pages 7576–7581. IEEE, Piscataway, N.J.

Schutz, H. A. (1988). The role of MAP in factory integration. *IEEE Transactions on Industrial Electronics*, 35(1):6–12, doi:10.1109/41.3056.

Seilonen, I. (2006). *An Extended Process Automation System: An Approach based on a Multi-Agent System*. Dissertation, Helsinki University of Technology, Helsinki.

Seilonen, I., Pirttioja, T., Pakonen, A., Appelqvist, P., Halme, A., and Koskinen, K. (2005). Information Access and Control Operations in Multi-agent System Based Process Automation. In Hutchison, D., Kanade, T., Kittler, J., Kleinberg, J. M., Mattern, F., Mitchell, J. C., Naor, M., Nierstrasz, O., Pandu Rangan, C., Steffen, B., Sudan, M., Terzopoulos, D., Tygar, D., Vardi, M. Y., Weikum, G., Mařík, V., William Brennan, R., and Pěchoucek, M., editors, *Holonic and Multi-Agent Systems for Manufacturing*, volume 3593 of *Lecture notes in computer science*, pages 144–153. Springer Berlin Heidelberg, Berlin, Heidelberg.

Shen, W. and Norrie, D. H. (1999). Agent-Based Systems for Intelligent Manufacturing: A State-of-the-Art Survey. *Knowledge and Information Systems*, 1(2):129–156, doi:10.1007/BF03325096.

Siemens (2005). Industrial Ethernet Products. www.siemens.com/profinet.

Siemens (2007). PROFINET – Der Industrial Ethernet Standard für die Automatisierung.

Silva, O., Garcia, A., and Lucena, C. (2003). The Reflective Blackboard Pattern: Architecting Large Multi-agent Systems. In Goos, G., Hartmanis, J., van Leeuwen, J., Garcia, A., Lucena, C., Zambonelli, F., Omicini, A., and Castro, J., editors, *Software Engineering for Large-Scale Multi-Agent Systems*, volume 2603 of *Lecture notes in computer science*, pages 73–93. Springer Berlin Heidelberg, Berlin, Heidelberg.

Skeie, T., Johannessen, S., and Holmeide, O. (2006). Timeliness of Real-Time IP Communication in Switched Industrial Ethernet Networks. *IEEE Transactions on Industrial Informatics*, 2(1):25–39, doi:10.1109/TII.2006.869934.

SMB Smart Grid Strategic Group (2010). *IEC Smart Grid Standardization Roadmap*. SG3.

Smith (1980). The Contract Net Protocol: High-Level Communication and Control in a Distributed Problem Solver. *IEEE Transactions on Computers*, C-29(12):1104–1113, doi:10.1109/TC.1980.1675516.

Spinnarke, S. (13.07.2016). OPC UA wird (neben anderen) Industrie 4.0-Standard. https://www.produktion.de/trends-innovationen/opc-ua-wird-neben-anderen-industrie-4-0-standard-334.html.

Spurgeon, C. E. and Zimmerman, J. (2014). *Ethernet: The Definitive Guide*. O'Reilly Media.

Sucic, S., Bony, B., Guise, L., Jammes, F., and Marusic, A. (2012). Integrating DPWS and OPC UA device-level SOA features into IEC 61850 applications. In *IECON 2012 - 38th Annual Conference of IEEE Industrial Electronics*, pages 5773–5778. doi:10.1109/IECON.2012.6389041.

Sucic, S., Havelka, J., and Dragicevic, T. (2014). A device-level service-oriented middleware platform for self-manageable DC microgrid applications utilizing semantic-enabled distributed energy resources. *Inter-*

national Journal of Electrical Power & Energy Systems, 54:576–588, doi:10.1016/j.ijepes.2013.08.013.

Sun, J., Rahman, M., Wong, Y., and Hong, G. (2004). Multiclassification of tool wear with support vector machine by manufacturing loss consideration. *International Journal of Machine Tools and Manufacture*, 44(11):1179–1187, doi:10.1016/j.ijmachtools.2004.04.003.

Tan, V., Yoo, D., Shin, J., and Yi, M. (2009). A Multiagent System for Hierarchical Control and Monitoring. *Journal of Universal Computer Science*, 15(13):2485–2505.

Tan, V., Yoo, D., and Yi, M. (2008). A Multiagent-System Framework for Hierarchical Control and Monitoring of Complex Process Control Systems. In Bui, T. D., Ho, T. V., and Ha, Q. T., editors, *Intelligent Agents and Multi-Agent Systems*, volume 5357 of *Lecture notes in computer science*, pages 381–388. Springer Berlin Heidelberg, Berlin, Heidelberg.

Thiel, K., Meyer, H., and Fuchs, F. (2010). *MES - Grundlage der Produktion von morgen: effektive Wertschöpfung durch die Einführung von Manufacturing Execution Systems*. Oldenbourg, München.

Thomesse, J. P. (1999). Fieldbuses and interoperability. *Control Engineering Practice*, (7):81–94.

Tidwell, D. (2000). Web services—the web's next revolution: IBM Tutorial. http://ibm.com/developerworks/.

Tobon-Mejia, D. A., Medjaher, K., Zerhouni, N., and Tripot, G. (2012). A Data-Driven Failure Prognostics Method Based on Mixture of Gaussians Hidden Markov Models. *IEEE Transactions on Reliability*, 61(2):491–503, doi:10.1109/TR.2012.2194177.

Trentesaux, D. (2009). Distributed control of production systems. *Engineering Applications of Artificial Intelligence*, 22(7):971–978, doi:10.1016/j.engappai.2009.05.001.

Twinoaks Computing Inc. (2011). What can DDS do for you? Learn how dynamic publish-subscribe messaging can improve the flexibility and scalability of your applications.

Ueda, K. (1993). A genetic approach toward future manufacturing systems. *Conference on Flexible Manufacturing Systems, Past, Present-Future*.

Ulewicz, S., Schütz, D., and Vogel-Heuser, B. (2012). Design, implementation and evaluation of a hybrid approach for software agents in automation. In *2012 IEEE 17th Conference on Emerging Technologies & Factory Automation (ETFA 2012)*, pages 1–4. doi:10.1109/ETFA.2012.6489766.

Unified Automation (2015). OPC UA is Enhanced for Publish-Subscribe (Pub/Sub). https://www.unified-automation.com/news/news-details/article/1217-opc-ua-is-enhanced-for-publish-subscribe-pubsub.html.

Valckenaers, P. (1993). *Flexibility for Integrated Production Automation*. PhD thesis, Katholieke Universiteit Leuven, Leuven.

van Biljon, S. S. (2004). *Role of access to 'real-time' information in the survival of enterprises*. PhD thesis, Stellenbosch : Stellenbosch University. http://scholar.sun.ac.za/bitstream/10019.1/16506/1/Van

van Brussel, H., Valckenaers, P., and Bonneville, F. (1994). Programming, scheduling and control of flexible assembly systems. *Manufacturing Systems*, 23(1):25–26.

van Brussel, H., Wyns, J., Valckenaers, P., Bongaerts, L., and Peeters, P. (1998). Reference architecture for holonic manufacturing systems: PROSA. *Computers in Ind.*, 37(3):255–274, doi:10.1016/S0166-3615(98)00102-X.

VDI (2012). VDI 5600 – Manufacturing Execution Systems – MES.

Vogel-Heuser, B., Diedrich, C., and Broy, M. (2013). Anforderungen an CPS aus Sicht der Automatisierungstechnik. *at – Automatisierungstechnik*, 61(10), doi:10.1515/auto.2013.0061.

Vogel-Heuser, B., Kegel, G., Bender, K., and Wucherer, K. (2009). Global Information Architecture for Industrial Automation. *atp - Automatisierungstechnische Praxis*, (1-2):108–115.

Vogel-Heuser, B., Schütz, D., Schöler, T., Pröll, S., Jeschke, S., Ewert, D., Niggemann, O., Windmann, S., Berger, U., and Lehmann, C. (2015). Agentenbasierte cyberphysische Produktionssysteme. *atp edition - Automatisierungstechnische Praxis*, 57(09):36, doi:10.17560/atp.v57i09.525.

Vyatkin, V. (2011). IEC 61499 as Enabler of Distributed and Intelligent Automation: State-of-the-Art Review. *IEEE Transactions on Industrial Informatics*, 7(4):768–781, doi:10.1109/TII.2011.2166785.

Wada, H., Sakuraba, Y., and Negishi, M. (2000). Machinery Control System using Autonomous Agents. *Proceedings of the 4th International Conference on Autonomous Agents*, Workshop, Agents Industry.

Warnecke, H.-J. (1992). *Die Fraktale Fabrik: Revolution der Unternehmenskultur*. Springer Berlin Heidelberg, Berlin, Heidelberg and s.l. doi:10.1007/978-3-662-06647-8. http://dx.doi.org/10.1007/978-3-662-06647-8.

Weinlaender, M. (2015). A New Auto-ID Integration Standard Could Play a Big IoT Role. http://www.rfidjournal.com/articles/pdf?12589.

Welber, I. (1987). Factory of the Future. *Control Systems Magazine, IEEE*, 7(2):20–22.

Wellenreuther, G. and Zastrow, D. (2011). *Automatisieren mit SPS: Theorie und Praxis ; Programmierung: DIN EN 61131-3, STEP 7, CoDeSys, Entwurfsverfahren, Bausteinbibliotheken ; Applikationen: Steuerungen, Regelungen, Antriebe, Safety ; Kommunikation: AS-i-Bus, PROFIBUS, Ethernet-TCP/IP, PROFINET, Web-Technologien, OPC ; mit 106 Steuerungsbeispielen und 7 Projektierungen*. Studium. Vieweg + Teubner, Wiesbaden, 5., überarb. und erw. aufl edition.

Weyrich, M., Diedrich, C., Fay, A., Wollschlaeger, M., Kowalewski, S., and Vogel-Heuser, B. (2014). Industrie 4.0 am Beispiel einer Verbundanlage: Aspekte der Modellierung und dezentralen Architektur. In VDI Wissensforum GmbH, editor, *Automation 2014*.

Wichmann, A. (07.06.2016). Industrial Security und Industrie 4.0: Sicherheitsanalyse von OPC UA.

Wiendahl, H.-P., ElMaraghy, H. A., Nyhuis, P., Zäh, M. F., Wiendahl, H.-H., Duffie, N., and Brieke, M. (2007). Changeable Manufacturing - Classification, Design and Operation. *CIRP Annals - Manufacturing Technology*, 56(2):783–809, doi:10.1016/j.cirp.2007.10.003.

Witsch, M. and Vogel-Heuser, B. (2012). Towards a Formal Specification Framework for Manufacturing Execution Systems. *IEEE Transactions on Industrial Informatics*, 8(2):311–320, doi:10.1109/TII.2012.2186585.

Wood, G. (1987). Survey of LANs and standards. *Computer Standards & Interfaces*, 6(1):27–36, doi:10.1016/0920-5489(87)90042-0.

Wooldridge, M. J. (2009). *An introduction to multiagent systems*. John Wiley & Sons, Chichester, U.K., 2nd ed. edition.

Wooldridge, M. J. and Jennings, N. (1998). *Agent technology: Foundations, applications, and markets*. Springer, Berlin and London.

World Economics Forum (2009). *Accelerating Smart Grid Investments*. World Economics Forum.

Wuest, T., Weimer, D., Irgens, C., and Thoben, K.-D. (2016). Machine learning in manufacturing: Advantages, challenges, and applications. *Production & Manufacturing Research*, 4(1):23–45, doi:10.1080/21693277.2016.1192517.

Wyns, J., van Brussel, H., Valckenaers, P., and Bongaerts, L. (1996). Workstation architecture in holonic manufacturing systems. *Proceedings of the 28th CIRP Internatinal Seminar on Manufacturing Systems*, Johannesburg, South Africa, Rand Africaans University:220–231.

Zhong, J., Yang, Z., and Wong, S. F. (2010). Machine condition monitoring and fault diagnosis based on support vector machine. In *2010 IEEE International Conference on Industrial Engineering and Engineering Management*, pages 2228–2233. IEEE, doi:10.1109/IEEM.2010.5674594.

Appendices

Appendices

A Protocols for Production Automation

A.1 History and Background on Fieldbus Systems in Modern Manufacturing

Although the term fieldbus originally appeared about 30 years ago in the middle of the 1980s, the technology behind these networking systems on the field level is much older (Sauter, 2005). The technological fundamentals of these communication technologies go back to first data transmission protocols used for the exchange of information through telephone lines that accordingly had to bridge long distances. The first standards of these "telex" networks were characterized by protocols such as V.21, X.21, X.25 and SS7. The advancement of these fundamental protocols finally resulted in the development of standards like the General Purpose Interface Bus (GPIB) (IEEE-488) characterized by parallel data and control lines. In further steps, networking technologies based on point-to-point connections were established that were not only able to bridge longer distances, but also to enable "multidrop" of data. The development of the first protocol that could deal with bus systems that are capable of establishing more than two connected instances and at the same time serving stable signal processing and noise resistance finally led to the establishment of the "RS-485" bus system that is still present in many fieldbus technologies today. (Sauter, 2010)

The original motivation to initiate systems in terms of modern fieldbus environments was to replace point-to-point connections between process control computers and actuators, which were before primarily characterized by star topology networking systems. The fieldbus systems intended to connect all devices within one automation system via a single serial bus (Sauter, 2010). The advantages of such an approach compared to prior configurations especially consist in an increasing flexibility and modularity in installing and extending industrial communication systems, and also during commissioning and maintenance (Thomesse, 1999). One factor of the

© Springer Fachmedien Wiesbaden GmbH, part of Springer Nature 2019
M. Hoffmann, *Smart Agents for the Industry 4.0*,
https://doi.org/10.1007/978-3-658-27742-0

access that fieldbus systems have had in these years is characterized through the modeling nature of fieldbus systems, which were the first attempt to model distributed systems not only from the networking point-of-view, but moreover from the application's perspective. Starting at the process and factory automation industry, fieldbus technologies were eventually further deployed into several other automation domains such as building and home automation, machine building, automotive and railway applications [...] (Sauter, 2010).

The first structured attempts of integrating field-level automation systems with the hierarchical organization introduced by the automation pyramid were strongly influenced by the International Organization for Standardization (ISO)/OSI model (Sauter, 2010). The reference architecture imposed by this model, which is still the starting point for the development of complex protocol standards, was the basis for the definition of the Manufacturing Automation Protocol (MAP) that was born in the course of the CIM endeavors (Sauter, 2010; Thomesse, 1999). MAP was intended to serve as a protocol for comprehensive control of industrial processes, but finally turned out to be too complex for a successful usage in automation practice (Schutz, 1988). In the following, many standardizations and fieldbus technologies had been brought to the street, e.g. the Military Standard 1533 bus, which was the first bus system to incorporate serial transmission of control and data information over the same line as well as the introduction of master-slave configurations with integrated controllers (Sauter, 2010). For industrial purposes, the Controller Area Network (CAN) fieldbus finally evolved and turned out to represent some kind of starting point for many proprietary fieldbus systems in the future.

Many of the protocol definitions that appeared during these standardization attempts had no real future in terms of sustainable advancements, as the underlying technological definitions focused more on specific problems in certain application domains than providing solutions to a broader field of action. Thus in the course of further developments, it became eventually clear that a reliable advancement of the fieldbus technology itself, addressing long-lasting communication needs in industrial automation, could only be successful if they were driven by "open" systems and communities (Sauter, 2010). Finally, the need for company independent standards turned out to be the most logical step for ensuring an extensive dissemination of these systems into the industrial reality of automated processes.

The first basic requirement of these standardization attempts was to ensure vendor independence, hence enabling a further increase of interoperability between devices of different vendors and automation hardware suppliers. For this purpose, so-called "companion specifications" – or in other words: companion standards, user layer, or networking standards – were carried out that firstly defined minimal data sets for certain applications domains, and at the same time introducing data syntax and semantics beyond the definitions of the OSI model (Sauter, 2005, 2010; Thomesse, 1999). The decisive step towards such an international standard was finally driven by a committee of the IEC, called SC65C, which finally came up with a standardization proposal for a "universally accepted fieldbus standard for factory and process automation" (Sauter, 2010). The two most promising approaches were the PROcess FIeld BUS (PROFIBUS) and the Factory Instrumentation Protocol (FIP) standards (Schutz, 1988; Felser, 2005), finally leading to the first standardization in the fieldbus sector after 14 years of technical and also politically motivated struggles. These multiprotocol standards, which contain many other protocol families besides the two mentioned, were introduced with the IEC 61158 and the IEC 61784 (DIN EN 61158-1, 2015; Felser, 2002)

Despite the development of various different fieldbus standards for the automation domain, the scope of these proprietary protocols turned out to be widely similar. Based on the nature of applications that had to be managed by the first fieldbus systems, the requirements of hardware connections on the shop floor were significantly different from those in LAN. Opposing the needs of office computer networks regarding the transfer of large data amount, fieldbus systems were more oriented to the requirements of manufacturing environments, in which comparatively small data packages needed to be exchanged, however on a high rate and with reliable delivery. Hence, real-time capabilities were the crucial element of those systems (Sauter, 2010). This guaranteed real-time data exchange ensures timely delivery of process-relevant information to the field devices that need to perform certain automation tasks and accordingly have to receive their control input right at that instance.

A typical fieldbus-based automation infrastructure is visualized in Figure A.1. Within these conventional rigidly organized hierarchies, there is usually a PLC that delivers control input variables and automation programs to the field devices and at the same time receives vital data from the devices through various I/O ports. In most of these applications, the PLC is placed in the middle layer between the automated systems and an interface layer that is usually represented by an HMI located at the top of the information hierarchy. These user access interfaces are finally able to interconnect machinery information from the shop floor with visualizations accessible to humans, e.g. maintenance workers in the field.

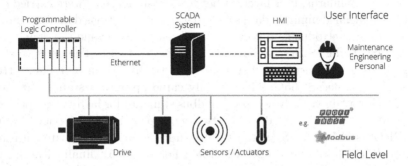

Figure A.1: Traditional hierarchical organization of a fieldbus system

As visualized in the figure, the production environment is basically divided into two sections, the lower level, which is also referred to as the "field level" or the "shop floor", and on the other side, the according visualization or user interaction layer that is intended to reflect the system states of the field devices. The interface between these dedicated systems is usually characterized by control systems such as PLCs and SCADA systems that are somehow present in both of the two worlds. The separation of these "two worlds" is still obvious as there is no natural interconnection between the tightly coupled systems in the field and possible interaction interfaces on the top. In this context, the upper level of the configuration shown in Figure A.1 reveals that its system composition rather concentrates on (passive) observation of the processes going on in the field than on interaction with the system. This degradation of a human interaction layer to some sort of information consumer without determined possibilities for a backflow of information demonstrates the weakness of such topology.

Hence, the major issues by setting up the automation environment in terms of fieldbus technologies consist in their rigid, hierarchical organization that complicate any adaptation or changes in the process configuration. This makes it – especially dynamically changing system environments – hard to set-up and to maintain. Another major drawback of coupling automation environments using fieldbuses is their lack in vertical interoperability, which can only be reached – if at all – at the expense of lowered security restrictions due to specialized, tailored connections to high-level information systems or components.

Another difficulty regarding fieldbus systems consists in the variety of available technologies, proprietary interfaces and pseudo standards. As already mentioned, even the standardizations in terms of fieldbus systems contain a vast amount of different protocol families that far from interoperable in a heterogeneous environment. Figure A.2 offers a glance at the huge number and contains only the most important fieldbus systems and their enabling technologies in terms of industrial automation, according to Sauter (2010). Even the international standards that attempt to align the different technologies on a common basis, are characterized by a vast number of completely disparate standards characterized by inhomogeneous system environments. The most dense area of known fieldbus standardizations are in fact located within the industrial and process section.

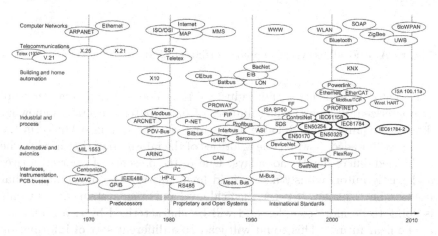

Figure A.2: Selection of important fieldbus systems and technologies (Sauter, 2010)

Even the standards that are known as most advanced within this industry sector do mostly not interoperate in the same automation environment without adapters or other externally deployed processors. This summery of emerged systems around fieldbus applications shows, how difficult it is to even oversee the market landscape of such protocols, despite understanding or even incorporating them in a unified system environment. All these fieldbus systems function within different topologies and by means of various exchange mechanism. A good overview gives Sauter (2010) as shown in Figure A.3:

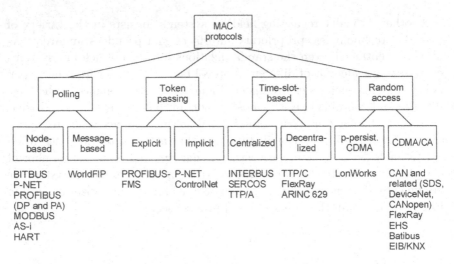

Figure A.3: Fieldbus system topologies (Sauter, 2010)

Recent developments within industrial shop floor environments that are characterized by an exploding number of data sources generate somewhat different requirements. Reliable and near real-time system behavior in the field is still needed, however the priorities of the information that needs to be exchanged differs slightly nowadays. Due to the presence of smarter devices and partly autonomously acting field level entities, the information that needs to be exchanged between the field level and other system environments is going to be more complex and will be characterized by rich semantics in the near future. This trend will lead to a different sort of information that contains more likely information about what to do rather then the exact steps to be performed by a field device. This shift in communication contents finally reduces the need for hard real-time signals in the shop floor while requiring a more complex communication between field devices and fieldbus systems.

Hence, especially in recent years, the technological background of system configurations on the field level have been significantly changing, as the network technologies that were originally designed only for office information technology have gained access to the lowest automation layers of the factory. In the course of these developments, the boundaries between fieldbus systems on the one hand and office environment networks have become blurry (Sauter, 2010). The reason for the merging of these networking systems lies in the requirement shift mentioned above that are not anymore solely characterized by hard real-time capabilities and fixed data rates focusing on proprietary, isolated protocols and networking solutions. However, despite the advancing merge of automation and IT networking systems, there is still one fundamental difference between these technologies. While LAN communication technologies in office environments are usually outdated or replaced in cycles of about three years, networking environments in automation and field level applications are still designed for time scales of more than ten years, commonly also up to 20 or even 25-30 years. Hence, the pace of technological advancements taking place within Information Technology (IT) cannot by kept up by the developments in automation networks. Therefore, despite the technological evolution for fieldbus systems, backward compatibility with currently running systems is always a crucial part that needs to be considered in further developments (Sauter, 2010). These compatibility requirements play a major role in the course of introducing new technologies and system environments by means of Industrial Ethernet based approaches, which will be examined as follows.

A.2 Basics and Historical Background of Industrial Ethernet

Due to the popularity of standard Ethernet (IEEE, 1980) the technology of exchanging information through broadband cable in LAN has been further developed for the usage at field level of factories supporting automation under real-time conditions (Jasperneite and Feld, 2005). This trend is a logical step, as the usage of LAN networks is based on the standardized Internet protocol (TCP/IP). Networking environments equipped with standardized Internet communication are characterized by constantly falling prices for the technical hardware and equipment, high bandwidth availability, established and reliable switching technology, and a rising number of product that meet the technical requirements for Ethernet-based networks (Phoenix Contact, 2005; Siemens, 2005). Other intrinsic capabilities of Ethernet are communication aspects such as data transport, network management, addressing, redundancy, discovery, security and safety (Prytz, 2008).

It is due to these advantages that Ethernet had attracted the interest of researchers soon after its establishment, also for the usage in industrial environments. At that time, however, the lack of capabilities in terms of real-time behavior was too obvious to be ignored and consequently led to serious obstacles when it came to setting-up such networking systems on the shop floor (Sauter, 2010). Soon, a number of attempts had been carried out trying to undermine this problem, e.g. by introducing traffic smoothing (LoBello et al., 2005). A breakthrough in terms of increasing the attractiveness of Industrial Ethernet for the usage in field level applications was introduced with the development of switching and full-duplex technology (Skeie et al., 2006). This step towards a better performing Ethernet has finally convinced industrial key players of their relevance for automation systems. As a matter of fact, many device vendors are nowadays representing the main contributors of the Industrial Ethernet movement (Felser, 2005). Although these improvements point into the right directions, the research around enabling real-time behavior based on Industrial Ethernet system still leave room for further improvement (Sauter, 2010; Decotignie, 2005).

A widespread utilization of Ethernet-based systems in industrial automation, however, had not been achieved already as the introduction of such system into the technical field layer was previously opposed due to the lack of real-time capabilities (Felser, 2005). In the meantime, several initiatives towards reaching near-real-time functionalities in Industrial Ethernet had been pursued, however, traditional fieldbus systems still play a major role in the automation domain (Jasperneite and Feld, 2005). Recent studies that concentrate on the adoption of Industrial Ethernet in automation have shown that the pace of transition increased in the last years (ARC Advisory Group, 2014). Thus, due to the functionality and investment protection, the switch from conventional fieldbus systems to Industrial Ethernet solution is a rather gradual process than a clear-cut transition (Jasperneite and Feld, 2005). An overview of the recent Industrial Ethernet standards and/or systems that focus on the mentioned evaluation benchmarks is given in Table A.1, according to Danielis et al. (2014).

Some rather new Industrial Ethernet that try to tackle these challenges have been introduced by means of Ethernet for Control Automation Technology (EtherCAT) and PROFINET IRT (Schumacher et al., 2008). They are both characterized as hardware-augmented Real-Time Ethernet (RTE) protocols, although this term "real-time" as it is used in the context of RTE does not comply to the common definition of that term as "real-time" with regard to Industrial Ethernet characterizes an intended communication process to be reliably finished in a determined time range. In practice, a near real-time behavior of Industrial Ethernet systems is achieved by adding real-time specific frames to the standard Ethernet storage area. An example for

Table A.1: Common Industrial Ethernet standards for DCS

Industrial Ethernet	Application domain / Explicit characteristic	Vendor	Delivery Time [ms]
PROFINET	Industrial manufacturing	Profibus & Profinet International	1
Ether-CAT	hard and soft real-time requirements in automation systems	Beckhoff Automation	0.15
Mod-busTCP	unlimited scalability and no special hardware requirements	vendor neutral	1-15
Ethernet Power-link	contains SPoF reliability	Ethernet POWERLINK Standardization Group (EPSG)	0.4
Ether-net/IP	relies on IEEE 1588 switches	Allen-Bradley/Rockwell Automation	0.13
SERCOS III	extremely low-delivery time	Bosch	0.0398

an integration of these additional time frames based on the PROFINET protocol is depicted in Figure A.4 (Siemens, 2007).

By adding PROFINET specific frames as shown in the figure, the delivery of information can be determined more precisely in advance. Each data package, i.e. payload information, is characterized by a distinguished Frame ID allowing for a specific identification of the information. The payload or "Process Data" is limited to an upper limit of Bytes. Hence, the data amount sent through each package and its according capacity and time consumption can be anticipated in advance. Some additional information about the Industrial Ethernet system status is propagated with every package by means of "Status Information". This complement, provided for the Industrial Ethernet standard intends to reduce the drawbacks that

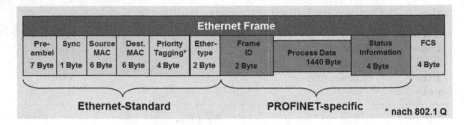

Figure A.4: Real-time specific components of the PROFINET Industrial Ethernet standard (Siemens, 2007)

standard Ethernet had to face when applied in industrial environments, in which package losses or late delivery can have a tremendous impact on the process.

As soon as Industrial Ethernet solutions in production environments will be realized, an interconnection between legacy systems like fieldbus systems and the Industrial Ethernet networking environment need to be realized in most cases. The practical implementation of such real-time information exchange between fieldbus systems and Industrial Ethernet systems is performed through some dedicated communication channel added to standard Ethernet. This channel is used for real-time critical payload as pointed out in Figure A.5. Thus, the information exchanged between the fieldbus environment and the PROFINET implementation is divided into non real-time critical information, such as configuration parameters or diagnostics information and the information that relies on timely delivery in terms of real-time referred to as the payload in the visualization. Both types of information are combined to be propagates through the Industrial Ethernet networking environment.

In order to allocate the development of EtherCAT and PROFINET IRT in the outline of Industrial Ethernet, the historical evolution of major Industrial Ethernet standards starting at the fieldbus systems are depicted in Figure A.6.

Other important well-established Industrial Ethernet systems are *Modbus-TCP, Ethernet Powerlink, TCnet, TCEthernet, CC-Link IE Field, EtherNet/IP, Sercos III (Bosch)*. A further, deep examination of the standards is provided by Danielis et al. (2014).

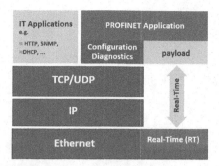

Figure A.5: Real-time information exchange in Ethernet systems

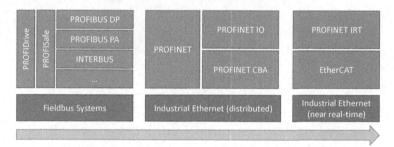

Figure A.6: Historical evolution of modern Industrial Ethernet protocols

B Architecture and Technical Realization of the MAS

The Appendix B.2.5 contains further details about architectural models and visualizations about multi-agent systems and their realization in terms of the OPC UA model definitions.

B.1 Multi-Agent System Architecture

The following figure shows a detailed class diagram of the original agent-based system architecture for Holonic Manufacturing Systems van Brussel et al. (1998). The PROSA reference architecture represents fundamental concepts for MAS and is still used in many applications.

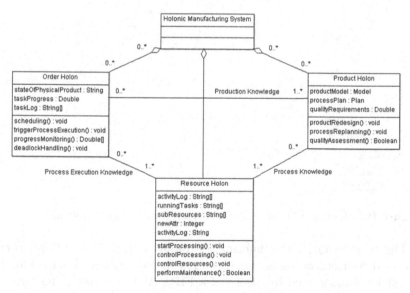

Figure B.1: Class diagram of the PROSA inspired HMS (van Brussel et al., 1998)

© Springer Fachmedien Wiesbaden GmbH, part of Springer Nature 2019
M. Hoffmann, *Smart Agents for the Industry 4.0*,
https://doi.org/10.1007/978-3-658-27742-0

B.2 OPC UA Based MAS

This section contains mappings of OPC UA based information models that are carried out in compliance with the FIPA agent and FIPA ACL standard. The *AbilityType* is further specified with regard to its filtering option for suitable agents.

B.2.1 GenericContentType Messages

The *GenericContentType* messages provide a simplistic approach to model ACL ontologies by means of an OPC UA information model.

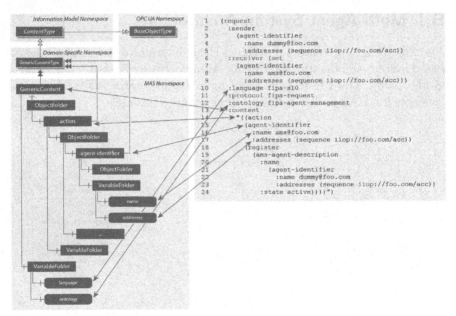

Figure B.2: Example Message of the *GenericContentType* mapping

The entry point of this information model is the *GenericContentType* object definition, which is able to represent any content description that might be standardized by means of a FIPA ACL message. The *Generic-Content* object that is instantiated using this type definition contains an *ObjectFolder* as well as a *VariableFolder*. The *ObjectFolder* is uses to model the actions of the ACL message, whereas the *VariablesFolder* maps the language and ontology definitions that are implied by the ACL message.

B.2.2 ContentType Subtype Mapping

Specialized subtypes of the *ContentType* definition are able to map complex message payloads on the OPC UA based approach for messages exchange between agents. In the following example a complex message is mapped onto the OPC UA information model.

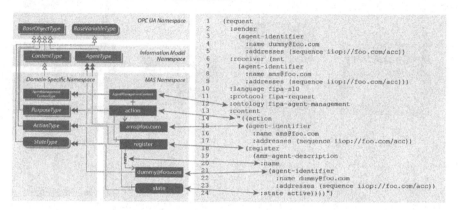

Figure B.3: Example Message of a specialized *ContentType* subtype mapping

Figure B.3 shows the modeling of an OPC UA based FIPA ACL compliant message in terms of an *AgentManagementContentType* instance. By instantiating the management content and the required sub types of the messages content (*PurposeType*, *ActionType* and *StateType*) the semantics of the message can be mapped by making use of the OPC UA information model shown on the left.

B.2.3 AbilityType Filtering Feature

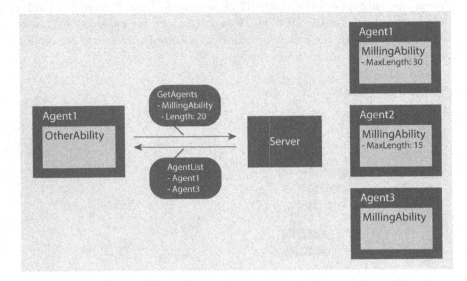

Figure B.4: *AbilityType* allowing parametrized requests from suitable agents

B.2.4 ERP Model Extension

Open Source ERP Systems

Table B.1: List of open-source ERPs

Name	Platform and technology	Software license	Last stable release
Adaxa Suite	Java	GPL	
Adempiere	Java	GPL	2015 (3.8 LTS)
Apache OFBiz	Java	Apache License 2.0	2016 (13.07.03)
Scipio ERP	Java	Apache Lic. 2.0/Commercial	2016 (1.14.1)
Compiere	Java	GPL/ Commercial	2010 (3.3.0)

Dolibarr	JavaScript, PHP, MySQL or Postgr-eSQL	GPLv3	2016 (3.8.3)
Epesi	PHP, MySQL	MIT license	2016 (1.8.0)
ERP5	Python, JavaScript, Zope, or MySQL	GPL	2014 (5.5)
ERPNEXT	Python, JavaScript, MariaDB	GPL	7 April 2016 (6.27.8)
GNU Enterprise	Python	GPLv3	2010
HeliumV	Java	AGPL	?
iDempiere	Java	GPLv2	2015 (3.1)
ino erp	PHP, JavaScript, MySQL/Oracle 12c	MPL	2016 (0.5.1)
JFire	Java, Eclipse	LGPL	2011 (1.2.0)
Kuali			
LedgerSMB	Perl, PostgreSQL	GPL	2016 (1.4.33)
Openbravo	Java, PostgreSQL, Oracle	OBPL1	2013 (3.0)
Odoo	Python, Javascript, PostgreSQL	LGPLv3/ Commer-cial	2016 (10.0)
Phreedom	PHP, Javascript, MySQL	GPLv3	2012 (3.4)
Postbooks	C++, JavaScript, PostgreSQL	CPAL	2014 (4.5.0)
SQL-Ledger	Perl, PostgreSQL	GPL	2016 (3.2.1)
Tryton	Python, GTK+, JavaScript	GPLv3	2016 (4.0)

ERP System Class Diagram

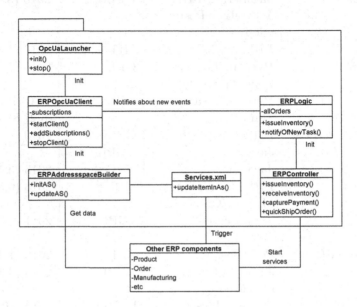

Figure B.5: Class Diagram of ERP system integration and components

B.2.5 Technical Setup of the Demonstrator

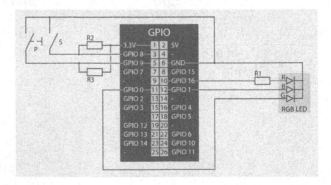

Figure B.6: Wiring diagram of the Raspberry Pi agent device for the OPC UA demonstrator

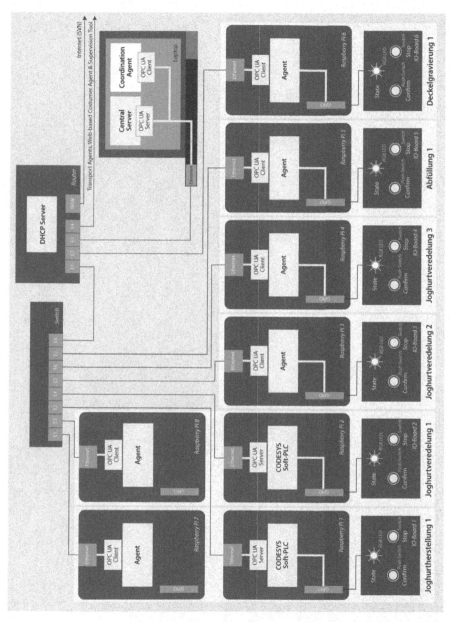

Figure B.7: Technical setup of the OPC UA demonstrator equipped with Raspberry Pi devices

C Machine Learning Results

C.1 Initial Prediction and Results of Genetic Algorithms

Figure C.1 show the results of the initial machine learning model after manual hyperparameter tuning without making use of an evolutionary approach.

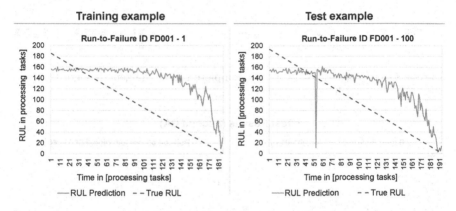

Figure C.1: Initial predication of the machine learning algorithm

The following figures show the results of the machine learning algorithms after to separate training sessions of genetic algorithms. As seen in both results, the run-to-failure prediction can be performed quite accurate.

© Springer Fachmedien Wiesbaden GmbH, part of Springer Nature 2019
M. Hoffmann, *Smart Agents for the Industry 4.0*,
https://doi.org/10.1007/978-3-658-27742-0

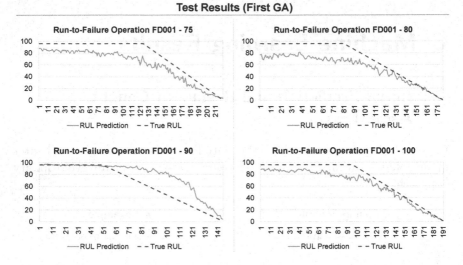

Figure C.2: Predication of the first complete run of the genetic algorithm approach

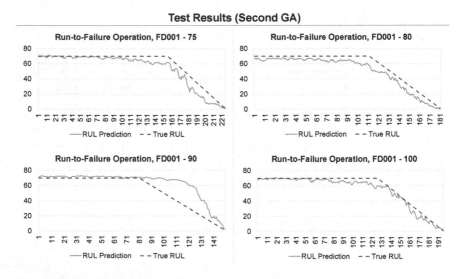

Figure C.3: Predication of the second complete run of the genetic algorithm approach

Printed in the United States
By Bookmasters